ISSN 1020-1203

Water Reports **24**



# RETHINKING THE APPROACH TO GROUNDWATER AND FOOD SECURITY

by
**Marcus Moench**
**Jacob Burke**
and
**Yarrow Moench**

FAO in collaboration with the
Institute for Social and Environmental Transition
Boulder, Colorado
United States of America

**FOOD AND AGRICULTURE ORGANIZATION OF THE UNITED NATIONS**
**Rome, 2003**

ISBN 92-5-104904-1

# Preface

This study attempts to re-frame the current thinking on groundwater development and the implications for food security. Groundwater is an important source for irrigated agriculture as it generally furnishes reliable and flexible inputs of water. To this extent, groundwater is instrumental in managing risk and optimizing food production. However, this reliance upon shallow aquifer systems for irrigation has turned to dependency. Competition for groundwater is intense both between neighbouring users and among economic sectors. This study highlights the role of adaptive strategies in dealing with aquifer management and indicates directions of research and management.

# Acknowledgements

This study is based on the work of Marcus Moench at the Institute of Social and Environmental Transition with assistance from Yarrow Moench. It has benefited from contributions and comments from Jacob Burke, Thierry Facon, Jean-Marc Faurès, Jippe Hoogeveen and Ase Elliason at the Land and Water Development Division of FAO.

# Contents

# List of boxes



# List of figures



# List of tables



# List of plates

# Core message

## Groundwater issues have important implications for food security, but effective responses require a change in perspective

Groundwater problems and the responses to them need to be viewed through a lens that is often different from existing management practice. While groundwater overdraft, pollution and quality problems contribute to food insecurity in many parts of the world, their implications for global food production levels cannot be estimated accurately. (In this paper, 'overdraft' refers to abstraction of groundwater that results in significant long-term declines in groundwater levels, without necessarily implying that the abstraction exceeds recharge.) Agricultural production based on groundwater is only one of numerous social, economic and environmental factors that determine food security. Furthermore, analysis of the links between groundwater overdraft and global food security yields little practical insight into how to address groundwater problems. Groundwater management is generally difficult to implement. Effective strategies are constrained by lack of data, limited technical capacities and, above all, lack of political and wider social support for standard regulatory or related interventions. Rapid demographic, economic and environmental changes in many regions often render such interventions inappropriate. By the time sufficient data and information are available to support management and the appropriate institutional frameworks are in place, conditions have often evolved to the point where initially envisioned management strategies have little chance of success. Change is an ongoing process and in many situations society is adapting rapidly to constraints and opportunities as they emerge.

## Adaptation is occurring and presents opportunities

Responses to emerging groundwater problems need to be identified on a more strategic basis than in the past. While integrated approaches to groundwater management in aquifers that are crucial to key uses (e.g. urban supply) may be feasible, the viability of establishing effective management systems over the large aquifers at risk in many countries is significantly lower. However, in many cases, people are responding to emerging groundwater problems through a variety of coping strategies. In addition, people are migrating or developing livelihood strategies that are not water intensive in some areas. These changes, the incentive for which is often unrelated to groundwater, could represent a major avenue for reducing the social consequences of groundwater overdraft and the pressure on the resource base. Adaptive strategies that build off existing social trends and responses to water problems may contain opportunities for effective responses that the standard water management lens misses.

## A strategic approach including adaptation is required

The limited groundwater management capacity (including the political capital needed to implement regulations and similar interventions) needs to be concentrated on a limited number

**Goundwater irrigation or thin linear aquifers in south Yemen**
*[J.J. Burke]*

of strategic, high-priority areas where it can be effective. In other regions, strategies that help people to adapt may prove more successful. Simple early-warning indicators that help guide such strategic approaches may be more important than detailed groundwater studies. Rather than attempting to identify global problems or apply a single broadly integrated management philosophy to all situations, there needs to be more focus on adapting groundwater response strategies to the array of factors that determine opportunities and constraints in specific situations (Plate 1).

# Chapter 1
# Introduction

The project on which this paper is based began as an attempt to evaluate the implications of groundwater overabstraction in key countries for food production and food security (China and India). The concept for the project emerged from discussions at the World Water Forum in The Hague in March 2000. There, much attention focused on a preliminary estimate by Postel (1999) that as much as 10 percent of global food production could be at risk from overabstraction. Postel's estimate was based on information collected from many sources and synthesized with input from a cross-section of experts. Although only a first cut, it generated substantial discussion and contributed considerably to recognition of both the role groundwater plays as a key resource for global food production and the threats now emerging as the resource is developed in a rapid and unregulated manner (Burke and Moench, 2000).

In formulating the current project, the aim was to move beyond Postel's numbers. By focusing on a few countries, the hope was to be able to locate substantial amounts of information from publicly available sources that would allow a more detailed evaluation of the extent to which groundwater overabstraction threatens food security. It was hoped that through a combination of contacts with experts in the focus countries and the rapidly expanding resources available through the World Wide Web (including the many organizations that have been developing and refining online access to their data) it would be possible to locate more basic data and information on groundwater than had been available to Postel. However, the real information available online is limited and generally consists of syntheses (or syntheses of syntheses) rather than basic data. Even where the area under groundwater irrigation is reported and professional estimates evaluated (FAO's Aquastat database), the actual dependency upon groundwater remains obscure (Burke, 2002). To date, it has only been possible to distinguish the contribution in crop yields and production from rainfed and irrigated agriculture (FAO, 2002a). Furthermore, as the search for information intensified (also through direct access to the basic data collected by the central and state governments in one of the focus countries, India), the less clear the equation linking groundwater and food security appeared. Nonetheless, the macro contribution of groundwater as a lead input for irrigated agriculture in all semi-arid and arid countries is evident. It is important to work towards a more refined understanding of this contribution in order to inform future management initiatives before the limits of pumping and groundwater quality are reached.

Available information indicates that groundwater overabstraction is a major emerging problem in many parts of the world. In many areas, overabstraction is severe and water levels are declining at rates of 1-3 m/year, a pattern documented by various researchers (Moench, 1996; Postel, 1999; Seckler *et al.*, 1999; Shah *et al.*, 2000). However, the areas where water levels are actually falling on a continuous basis remains poorly documented, also in the data sets compiled by government groundwater departments working at local levels. Variability within and between regions is high. Furthermore, overabstraction estimates and water-level decline information represent only one component in the equation linking groundwater, food production and food security. Water-level fluctuations can have equally great impacts on access to water (and thus food production) as can overabstraction and long-term declines. Other important elements of the equation such as yields, extraction economics and the composition of demand for water are often poorly documented. Finally, the question of groundwater and food security is a product of interacting and dynamic systems. As the relevant hydrological, economic and agricultural systems are dynamic and often undergoing rapid change, historical data may be of limited usefulness for predicting future conditions. As a result, any estimate of the impact of groundwater overabstraction on food security would contain substantial uncertainty.

This paper focuses on the question of how well the interaction between groundwater and food security is understood and what types of information are essential to improve this understanding. It follows the definition of food security in the FAO terminology database (derived from the World Food Summit 1996) that: "*Food security exists when all people, at all times, have physical and economic access to sufficient, safe and nutritious food to meet their dietary needs and food preferences for an active and healthy life*". The paper explores issues such as data availability, the dynamic nature of systems and the interaction between systems. The overall question of groundwater and food security is introduced from the perspective of both resource availability for production and the larger question of entitlements and access. The paper then focuses on groundwater issues per se and on how the combination of data limitations and regional-temporal variability constrains accurate assessment. It then draws implications for the analysis of food security and for groundwater management. The paper concludes with a discussion of the potential ways for addressing concerns relating to groundwater and food security.

The results of the analysis suggest two avenues for future work. The first involves the development of a detailed research programme to harvest groundwater data directly from governments and other sources within key countries in order to develop a more complete picture of groundwater use and conditions. This type of picture is essential in order to develop more informed assessments of the implications of groundwater conditions for food security and to develop scientifically founded courses of action for managing the resource base. The second avenue for work focuses on the development of adaptive responses to water problems and of policy approaches that reflect and respond to uncertainty and change. Inherent limitations in the nature of scientific information in conjunction with the dynamic process of social and institutional change occurring in many parts of the world make this second avenue of work no less important than the first. The most effective approaches to addressing overabstraction problems may lie not in attempts to manage the resource base but in strategies that assist populations to lessen their dependence on groundwater or that allow them to adapt as water availability changes. Furthermore, experience with common-property resource management suggests that the ability to reach effective management agreements tends to be inversely proportional to the number of individual actors whose participation is essential. This can be resolved in the case of surface systems by designing management approaches as a series of nested organizations with defined boundaries in relation to their share of the resource (Ostrom, 1993). In locations such as India and China, there can be tens of thousands of individual farmers tapping shallow stratiform aquifers that underlie large areas. In such cases, a critical challenge is to define boundaries at a sufficiently small scale for nested institutions or 'cells' to function and replicate.

The above observation reflects the frequent tension between access equity and sustainability that is one of the many contradictions inherent in most attempts to manage natural resources. It also reflects the dynamic nature of social change that is ongoing in many parts of the world. In the past decade, many farmers have indicated that, while they are aware of groundwater problems,

PLATE 1
**Subsidized groundwater pumping in the Zhob Valley, Baluchistan**
*[J. J. Burke]*

they are using the income generated through agriculture to invest in alternative livelihood strategies (education, businesses, etc). Microirrigation using low-head and low-cost technology, including drip and treadle pumps, has shown promise where local markets exist (e.g. near towns and cities) but much of this initiative is not captured in official national irrigation data. The focus of rural populations in many areas is changing and many individuals have more incentive to use the income from irrigated agriculture to build non-agricultural livelihoods. Helping them to make this transition could yield major benefits for the sustainability of groundwater resources while at the same time responding to their own perceived goals and interests as shaped by much larger processes of social change (Plate 1).

Contradictions are inherent in any robust, dynamic system. They reflect the checks, balances and points of duplication essential to respond to change, uncertainty and surprise. For FAO, this implies a need to approach groundwater through multiple lenses: management where that appears practical, adaptation where it appears less viable, and common sense in both situations.

# Chapter 2
# The links between groundwater and food security

The following quotations from D. Seckler, former Director of the International Water Management Institute (IWMI), are characteristic of statements by many experts and of the tenor of the discussions at the World Water Forum held in March 2000:

"*For most of modern history, the world's irrigated area grew faster than population, but since 1980 the irrigated area per person has declined and per capita cereal grain production has stagnated. The debate regarding the world's capacity to feed a growing population, brought to the fore in the writings of Malthus two centuries ago, continues. But the growing scarcity and competition for water add a new element to this debate over food security. ... In a growing number of countries and regions of the world, water has become the single most important constraint to increased food production*" (Seckler *et al.*, 1998).

"*Many of the most populous countries of the world - China, India, Pakistan, Mexico, and nearly all of the countries of the Middle East and North Africa - have literally been having a free ride over the past two or three decades by depleting their groundwater resources. The penalty of mismanagement of this valuable resource is now coming due, and it is no exaggeration to say that the results could be catastrophic for these countries, and given their importance, for the world as a whole*" (Seckler *et al.*, 1999).

Water scarcity is recognized increasingly as a global concern and, within that broad concern, more attention is focusing on emerging patterns of groundwater overexploitation and their implications for the availability of water to meet human and environmental needs. Food production and food security are among the more important of these needs. Reliable water supplies, particularly those from groundwater, are the lead input for increasing yields, reducing agricultural risk and stabilizing farm incomes. As a result, strong arguments can be made that access to groundwater plays an instrumental role in food security.

Water availability and reliability are linked closely to food security, but the equation linking water to food security is partial and the links are neither linear nor transparent. The full equation is a function of the interaction between water access, production economics and the wider network of entitlements that water users and others have within society. It cannot be assumed that a one-to-one relationship exists between access to reliable water supplies for irrigated agriculture and food security (Plate 2).

According to technical documents prepared for the World Food Summit in 1996, the main generally available indicator of food security is: "*per caput food consumption, measured at the national level by the average dietary energy supply (DES) in calories on the basis of national food balance sheets (FBS) and food supplies as national averages*" (FAO, 1996). The FAO definition of food security (above) that this paper follows does not focus on food production and physical availability alone; it also includes the critical dimension of access to available food supplies. Under the definition, food security often depends more on the ability of populations to purchase rather than produce food. This is because global and national food distribution systems now frequently negate the impact of local production problems on the availability of food in the market. As a result, the question of whether people have access to sufficient food when groundwater problems disrupt agricultural production depends heavily on whether they have access to a diverse array of alternative income sources or to reserve capital. It also depends on wider factors such as transportation systems and the ability of countries to purchase and distribute food available on global markets. All this implies that analysis of groundwater availability and reliability on a project or regional basis is by itself a poor indicator of the vulnerability of the global population to food insecurity (Figure 1).

Nonetheless, access to water, and particularly to highly reliable groundwater sources, does play an important role in food security in many cases.

**Plate 2**
**Pumping in Gujurat, India. One farmer + one pump = food security – or does it?**
*[M. Moench]*

**FIGURE 1**
**Number of undernourished in the developing world: observed and projected ranges compared with the World Food Summit target**

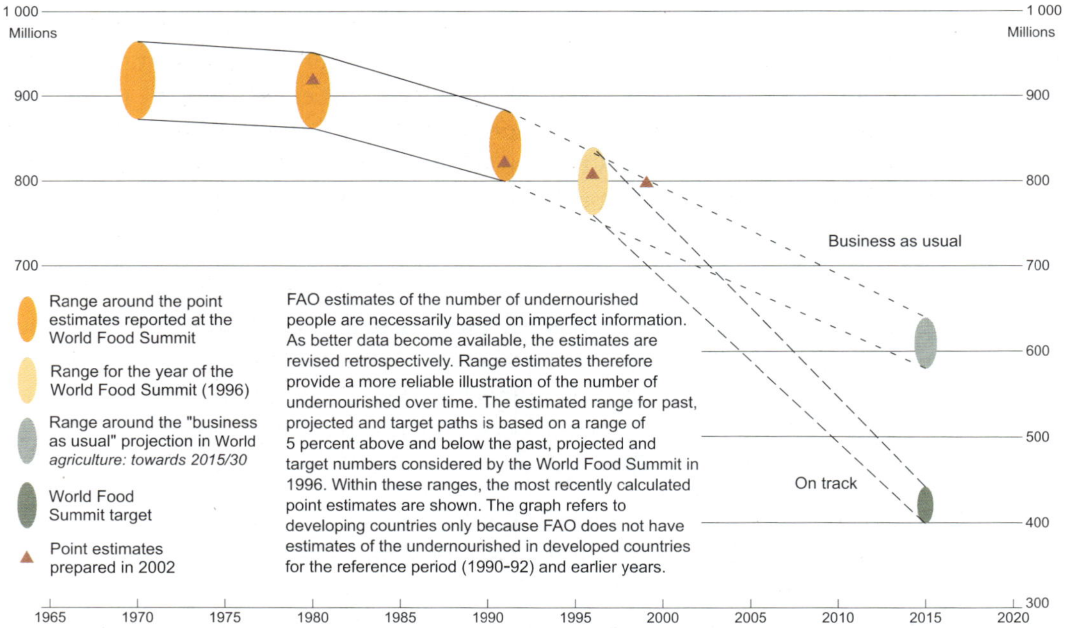


Source: The State of Food Insecurity in the World. FAO, 2002

Access to reliable sources of water reduces the production risk. Farm incomes at both micro (farm) and aggregate (regional) levels are buffered from the effects of precipitation variability, drought or general water scarcity conditions. As a result, access to reliable groundwater supplies can ensure the income flow needed to purchase food as well as playing a central role in food production. Furthermore, particularly in remote locations within developing countries, irrigated agriculture constitutes the sole source of income that is available to rural populations. As a result, there can be a direct link between water access and household or regional food security. However, this link is highly dependent on the specific situation. There is no inherent direct link between water and food security. While access to water is important in many situations, in other situations irrigated agriculture is only one of many income sources or available livelihood strategies. Consequently, while falling water levels, irrigation system deterioration, droughts and other direct indicators of water scarcity can serve as signals that food security may be threatened, the actual degree of threat will depend on a wide variety of context-specific factors. Water scarcity measures are warning signals, but they do not on their own indicate the emergence of food insecurity.

## Resource availability and production

The most direct and tangible link between groundwater conditions and food security is that of water availability to meet crop requirements. However, water availability in an aggregate sense has little meaning as crop production is heavily dependent on seasonal and interannual fluctuations in availability, including timing in relation to crop growth stages. Many crops are vulnerable to moisture stress at critical points in plant growth, and their yields can be reduced substantially even if adequate water supplies are available following periods of shortage (Perry and Narayanamurthy, 1998). For example, water stress at the flowering stage of maize can reduce yields by 60 percent even where water is adequate throughout the rest of the crop season (Seckler and Amarasinghe, 1999). Similar impacts on onions, tomatoes and rice have also been documented (Meinzen-Dick, 1996). In addition to the direct impact of water availability on crop growth, assured supplies are a major factor in inducing investment in other production inputs such as labour, fertilizers, improved seeds, and pesticides (Seckler and Amarasinghe, 1999; Kahnert and Levine, 1989). As a result, as the reliability of irrigation water supplies increases there is multiplier effect on yields. Taken with the inherent flexibility of groundwater abstraction (on demand, just in time), these characteristics of groundwater were a major contributor to the role of irrigation in the green revolution. Irrigated agriculture now contributes almost 40 percent of world food production from 17 percent of cultivated land (United Nations, 1997).

Expansion of irrigation was the 'lead' input driving yield increases during the green revolution of the 1960s-70s and subsequent decades. As the most reliable source of irrigation water, a source that can generally be tapped when and in the amounts needed, groundwater played a particularly major role. As Repetto (1994) comments: *"The Green Revolution has often been called a wheat revolution; it might also be called a tubewell revolution."* To this extent, this turnaround hinged upon high-value crops (with high crop-water budgets) and the ability to pay for energy costs (Plate 3).

Yields in groundwater-irrigated areas are higher (often double) compared to those in canal-irrigated areas (Shah, 1993; Meinzen-Dick 1996). In India, the groundwater-irrigated area accounts for about 50 percent of the total irrigated area and up to 80 percent of the country's total agricultural production may, in one form or another, be dependent on groundwater (Dains and Pawar, 1987). Similar patterns are also present in other countries. In China's Henan province,

**Plate 3**
**Extensive groundwater irrigation of wheat from large stratiform aquifers. Sadah, Yemen**
*[J.J. Burke]*

tubewells serve about 2 million ha, or 52 percent of irrigated lands (FAO, 1994). Parts of Mexico, including some of its most productive agricultural areas, are also heavily dependent on groundwater. The role of groundwater is equally important in industrialized countries. For example, Barraque (1998) estimates that: "*irrigation uses 80 percent of all water in Spain and 20 percent of that water comes from underground... The 20 percent, however, produces more than 40 percent of the cumulated economic value of Spanish crops.*" Recent findings from Andalusia, Spain, indicate that groundwater-irrigated agriculture is economically more than five times more productive (in terms of revenue per cubic metre) and generates more than three times the employment in comparison with surface-irrigated agriculture (Hernandez-Mora *et al.*, 1999). The role of groundwater is important not only through higher yields in normal water years. In an analysis of wheat cropping in the Negev Desert, Tsur (1990) estimated the 'stabilization value' (the value associated with the reliability of the water supply as opposed to just the value of the volume available of groundwater development) as being "*more than twice the benefit due to the increase in water supply*". In southern California, the United States of America, where surface water supplies are less variable than in the Negev Desert, the stabilization value in agriculture is as much as 50 percent of the total value of groundwater in some cases (Tsur, 1993). During the drought in California, the United States of America, in the early 1990s, economic impacts were minimal largely because farmers were able to switch from unreliable surface supplies to groundwater (Gleick and Nash, 1991). The value associated with the flexibility of pumped groundwater supplies has been a further boost to agricultural productivity as it has allowed intensification and diversification of agricultural production in otherwise inflexible surface-irrigation schemes.

However, the presence of groundwater irrigation alone cannot be given full credit for the increased yields documented around the world. It needs to be seen as part of a complementary and mutually reinforcing set of inputs. Groundwater availability has enabled farmers to invest in complementary inputs that, in combination, have increased crop yields substantially. As FAO (2002b) notes: "*the response of crop to fertilizer is higher where supply of irrigation water is assured compared to rainfed conditions.*" It is the reliability and flexibility of groundwater that allows farmers to take the risk of investing in fertilizer, but which also substantially increases their crop productivity. For example, fertilizer use in Pakistan is highest in areas supplied by both canals and tubewells and thus having a highly assured supply of irrigation water. The total nutrient application in these areas is 68.85 kg/ha compared to 29.0 kg/ha in rainfed areas (FAO, 2002b). These observations point to the dependency of crop yields on interactions within a dynamic agricultural system and the difficulty of isolating a single factor as the primary factor contributing to increased production.

Nonetheless, the information available indicates the critical role groundwater has played in agricultural production over recent decades. The relationship between assured supplies of irrigation water, increasing yields and food production is now under stress. According to Rosegrant and Ringler (1999): "*the growth rate in irrigated area declined from 2.16 percent/ year during 1967-82 to 1.46 percent in 1982-93. The decline was slower in developing countries, from 2.04 percent to 1.71 percent annually during the same periods.*" Yield increase rates are also declining, and projections indicate that this will continue in coming decades (Rosegrant and Ringler, 1999; FAO, 2000). Furthermore, in some local areas such as Sri Lanka and in the rice-wheat systems of India, Nepal, Pakistan and Bangladesh, yields have been stagnant for a number of years (Amarasinghe *et al.*, 1999; Ladha *et al.*, 2000).

Although stresses on water resources are increasing and there is a logical link between water scarcity and yield stagnation, causal relationships between emerging water problems, yields and food production vulnerability are not proven. According to Ladha *et al.* (2000), where yield stagnation is concerned: "*There is some evidence of declining partial or total factor productivity...The causes for the stagnation or decline are not well known, and may include changes in biochemical and physical composition of soil organic matter (SOM), a gradual decline in the supply of soil nutrients causing nutrient (macro and micro) imbalances due to inappropriate fertilizer applications, a scarcity of surface water and groundwater as well as poor water quality (salinity), and the buildup of pests, especially weeds such as Phalaris minor.*"

Furthermore, as Seckler and Amarasinghe (1999) note: "It is very difficult to project crop yields. … The international dataset does not distinguish between yields on irrigated and rain-fed area: they are just lumped together in average yields." Water is only one factor affecting crop yields. Data available at the global level do not provide much insight into the relationship between yields on irrigated and rainfed lands, or enable conclusions about yields on areas irrigated by groundwater or on areas where groundwater depletion is occurring. Recent evaluations of the implications of water scarcity for food security range from the optimistic to the pessimistic. For example, Brown (1999) contends that primarily because of impending water shortages in northern China, the country will have to import up to 370 million tonnes of grain per year to feed its population in 2025. This massive increase in imports could cause steep increases in cereal prices and disruption of the world market (Seckler *et al.*, 1999). On the other hand, analyses by FAO and the International Food Policy Research Institute (IFPRI) indicate that yield increases (rather than increases in cultivated area) will be the dominant factor underlying growth in cereal production in the coming decades and that, in aggregate, production increases will be sufficient to meet demand (Rosegrant and Ringler, 1999; FAO, 2002a). The FAO (2002a) report states that: "The overall lesson of the historical experience, which is probably also valid for the future, seems to be that the production system has so far had the capability of responding flexibly to meet increases in demand within reasonable limits." (Plate 4).

The core point in this discussion of the direct links between groundwater availability and food production is the role of interacting dynamic systems and the uncertainty inherent in predicting outcomes based on partial understanding of any one of them. The role that irrigation, particularly groundwater irrigation, has played in increasing yields is relatively clear. Whether or not emerging water problems are a significant factor underlying the declining rate of yield increases or represent a significant threat to overall production levels is less well documented.

While the potential nature of such connections is clear in concept, available data and other evidence are insufficient to test the conceptual relationships. While it is essential not to dismiss the implications of groundwater overabstraction and water scarcity for food production simply because data are insufficient to prove them, it is equally essential not to ignore the wide range of other factors that could be playing equal or greater roles. Therefore, the first part of the equation linking groundwater and food production is clouded even before investigation of the larger question of the role groundwater plays in entitlements and food security.

**Plate 4**
**The physical and economic limits to pumping, Eritrea**
*[J.J. Burke]*

## Entitlements and food security

Food security is a function of three factors: availability; stability; and the ability of individuals to obtain access to food. As Sen (1999) and others (Dreze *et al.*, 1995) have argued for famines in India, starvation is frequently due to the inability of individuals to purchase supplies that are readily available on the market and is not a function of food availability per se. The entitlement approach described by Sen "*views famines as economic disasters, not just as food crises.*" Sen indicates that the main interest in the entitlement approach probably lies in "*characterizing the nature and causes of the entitlement failures where such failures occur.*"

Sen's approach may have particular relevance for analysing the impact of emerging groundwater problems on food security. Studies in the late 1980s highlighted the critical role that access to water, particularly groundwater, plays in poverty alleviation (Chambers *et al.*, 1989). Reliable water supplies are a foundation that enables farmers to afford access to a wide range of development benefits (from food to education and health services) and can also enable farmers to diversify into other, often non-agricultural, income sources. These benefits are accessed through the improved yields enabled by the green revolution package of inputs. However, they carry a substantial risk because farmers must make investments in fertilizer, seed and other inputs in order to achieve them. These investments, which are often made on credit, will be lost if water supplies fail. Consequently, any decline in access to groundwater could have a major impact on the economic condition of small rural farmers. As Burke (2000) argues: "*the expansion of irrigated agriculture in the 20$^{th}$ century has de-coupled the water user from the inherent risk of exploiting both surface and groundwater resources. The apparent reliability of storage and conveyance infrastructure and the relative cheapness and flexibility of groundwater exploitation offered by mechanical drilling have sheltered the end user from natural hydrological risk.*" If substantial groundwater-level declines occur, short-term risk exposure may return to levels not encountered since the spread of irrigation. This risk is predominantly economic.

The economic dimension is also central to understanding the meaning of groundwater overextraction. Most discussions of groundwater overabstraction emphasize the distinction

**Plate 5**
**Tension between farmers and municipalit users. Taire, Yemen**
*[M. Moench]*

between economic depletion (i.e. falling water levels make further extraction uneconomic) and the actual dewatering of an aquifer. Large-scale aquifers are depleted in an economic sense (the physical limits to pumping and associated energy costs) long before there is any real threat of physical depletion. The Gangetic basin may have 6 000 m of saturated sediment, but only the top 100 m or so are economically accessible for irrigation. Furthermore, wells owned by small farmers are generally shallow. In the context of poverty and famine, falling water tables will tend to exclude those farmers who cannot afford the cost of deepening wells long before they affect water availability for wealthy farmers and other affluent users (Moench, 1992). Consequently, substantial declines in water levels are particularly likely to have a major economic impact on farmers with limited land and other resources. This impact will tend to be particularly pronounced during drought periods when large numbers of small farmers could simultaneously lose access to groundwater as their wells dry up. A more creeping problem would occur during non-drought periods as water-level declines undermined the economic position of small marginal farmers, forcing them onto already saturated unskilled agricultural and urban labour markets. The food security crisis in both these situations would be economic rather than related to foodgrain availability per se. Furthermore, whether there actually is a food security problem would depend as much on larger economic conditions (specifically the opportunities available to farmers transferring out of agriculture into other activities) as on groundwater availability and the economics of agriculture (Plate 5).

The question of the larger economic situation is particularly relevant in the context of global demographic and economic changes. Although the latest UN assessment indicates a substantial deceleration in world demographic growth rates, the absolute annual increments in the coming decades will continue to be large. According to FAO (2000): "*seventy-seven million persons are added to world population every year currently. The number will not have decreased much by 2015. Even by 2030, annual additions will still be 58 million.*" Ninety-eight percent of the increase between 1995 and 2020 will occur in the developing world with the largest absolute growth concentrated in Asia and the highest relative increases occurring in sub-Saharan Africa (Pinstrup-Anderson *et al.*, 1999). Population growth will be accompanied by significant changes in where people live. Historically, rural populations have dominated those living in urban areas. However, within the next 15 years, the urban population in developing countries is projected to surpass the rural population (Pinstrup-Anderson *et al.*, 1999). Furthermore, as populations urbanize, their aspirations and food-demand characteristics will change. Such changes are reflected in recent food trade and demand projections. Over the next 20 years, according to Rosegrant and Ringler (1999): "*Per capita food consumption of maize and coarse grains will decline as consumers shift to wheat and rice, livestock products, fruits and vegetables, and processed foods. The projected strong growth in meat consumption, in turn, will substantially increase cereal consumption as animal feed, particularly maize. Growth in cereal and meat consumption will be much slower in developed countries. These trends will lead to a strong increase in the importance of developing countries in global food markets: 82 percent of the projected increase in global cereal consumption, and nearly 90 percent of the increase in global meat demand between 1993 and 2020 will come from developing countries. Developing Asia will account for 48 percent of the increase in cereal consumption, and 63 percent of the increase in meat consumption.*"

To meet changing demand patterns, FAO and the IFPRI project substantial increases in world trade for food, particularly cereals and meat products (FAO, 2000; Rosegrant and Ringler, 1999). Cereal trade is projected to almost double and meat trade to triple by 2020. To date, traditional cereal exporters (North America, Australia, Argentina, Thailand, Western Europe and Viet Nam) have been able to meet sudden rises in demand in developing countries. However, Brown (1999) points out that grain exports by the principal exporting countries (accounting for 85 percent of world exports) have levelled off since 1980. There is debate as to whether developing countries will be able to meet food needs through trade. However, it could also be argued that the high population growth in some water-stressed developing countries (e.g. Jordan and Palestinian Authority) shows that food production is not a limit to food security.

Returning to the question of the link between groundwater conditions and food security, Sen's framework suggests that access to groundwater will continue to play a critical role in the network of entitlements that determine food security for rural agricultural populations. However, as populations migrate from rural to urban areas, direct access to groundwater for individuals will play less of a role. This is also the case where rural economies become less dependent on agriculture. Furthermore, the food security impact of groundwater-level declines on rural agricultural populations will depend as much on their ability to join the stream of permanent or temporary migrants to urban areas as on their ability to maintain economic livelihoods in rural areas. On an anecdotal level, this dynamic is evident in discussions with farmers in diverse conditions. For example, in many interviews with farmers in Gujarat, India, concerning groundwater overabstraction and the possibility of developing management systems, discussions have elicited the following type of response: "*Yes, I know falling water levels will drive me out of production in a few years, but why should I care? The income I am generating now is enabling my children to study for an engineering degree; we will not be here in five years' time.*" Farmers in the United States of America often express similar sentiments. A 'young' farmer in the San Luis Valley in Colorado, the United States of America is 65 years old; a wide set of economic and social factors has induced many young people to prefer a livelihood in the urban or non-agricultural economy. From a food security perspective, they have joined the half of the world's population that depends on global economic, production and distribution systems within which groundwater availability is only one element. In this sense, they no longer depend on direct individual access to local resources such as groundwater.

For urban residents and the increasing population not engaged in agriculture, food security is likely to become a function of distant production and distribution systems combined with the economic context individuals find themselves in. This implies that food security for many will be influenced at least as much by conditions in the wider economy as by factors such as groundwater conditions that affect agricultural economics and local agricultural production per se.

## Environmental data and environmental myth

The discussions above point to the complex nature of the interactions between groundwater conditions and a large number of factors in the wider global economic and demographic context that influence food security. Full analysis of these factors is beyond the scope of this paper. However, recognition of the complexity and identification of points of leverage within it is critical to any meaningful analysis of the implications of groundwater overabstraction for food security. The complexity is also central to identifying meaningful responses to emerging water problems. Many compelling analyses of environmental-social relationships have foundered on seemingly minor gaps in data or system descriptions.

The next section of this paper focuses on groundwater: how well the resource base and emerging overabstraction problems are understood and, beyond that, the implications of emerging problems for food production and security. However, before that, Boxes 1 and 2 present two cautionary examples that highlight the fundamental risk inherent in posing major global consequences where there is partial or weak scientific understanding and where the systems involved are complex.

The cautionary examples in Boxes 1 and 2 contain lessons that are central to the question of evaluating the impacts of groundwater overabstraction on food production and food security.

First, the database relevant to the question being asked was weak in both cases. This was natural enough in the 1870s-1880s when long-term rainfall records were unavailable and changes

**Box 1: Rain follows the plough**

In the last decades of the nineteenth century, throughout the western United States of America, settlers received land grants of 160 acres (about 65 ha) under the Homestead Act. The High Plains, an area of rolling grasslands between the Mississippi River and the Rocky Mountains were a focal point for settlement. Each year, the waves of settlers drove slightly further west, claimed land and broke the sod. In one decade, nearly 2 000 000 people settled on the Great Plains.[1]

"God speed the plow.... By this wonderful provision, which is only man's mastery over nature, the clouds are dispensing copious rains ... [the plow] is the instrument which separates civilization from savagery; and converts a desert into a farm or garden.... To be more concise, rain follows the plow". (Charles Dana Wilber)[2]

"Rain Follows the Plow" was a common headline on brochures promoting settlement. This statement was based on accepted scientific analysis of the day. Soil beneath the grasslands was rich and often very moist. Ploughing it would release substantial moisture. According to the theory, as more land was brought under cultivation, more moisture would be released. This would, in turn, contribute to cloud formation and ultimately cause rainfall to increase. The High Plains could be converted into a climate resembling the temperate moist areas of the east coast. This was 'science', an integrated theory grounded on an apparent understanding of the physical processes incorporated in the 'model'. Furthermore, the climate behaved as predicted. Unusually heavy rainfall in the 1870s and early 1880s, made the claims sound plausible. Rainfall in the High Plains appeared to increase as agricultural areas grew. The story declined as rainfall returned to the lower levels common throughout much of recent history. The endnote was the Dust Bowl, that great event that reshaped much of rural America in the 1930s.

Where did the analysis go wrong? First, although the processes were understood, at least at a gross level, the orders of magnitude on the flows involved were not. Ploughing may release moisture from the soil, but the amounts involved were far too small to affect the regional climate. Second, the 'model' neglected interactions with other systems. Regional and hemispheric wind patterns are such that any water released from ploughing soil in the High Plains is often transported huge distances before it reaches the ground as running water again. Rain did not follow the plough, it only appeared to.

[1] PBS, available at: http://www.pbs.org/weta/thewest/wpages/wpgs100/w171_002.htm.
[2] PBS, citing Charles Dana Weber, available at: http://www.pbs.org/weta/thewest/wpages/wpgs100.

**Box 2: Himalayan deforestation**

One of the most compelling recent environmental stories was that surrounding deforestation in the Nepal Himalayas. In a classic analysis, Eckholm (1976) painted a picture of environmental degradation in the hills having major regional consequences. The 'model' was clear. It envisioned a direct cause-and-effect relationship between population growth, mountain deforestation and lowland flooding. "As wood scarcity forces farmers to burn more dung for fuel, and to apply less to their fields, falling food output will necessitate the clearing for ever larger, ever steeper tracts of forest - intensifying the erosion and landslide hazards in the hills, and the siltation and flooding problems downstream in India and Bangladesh." (Eckholm, 1976). The interaction between population, food and fuel lay at the heart of deforestation problems. As forest cover declined, erosion and the speed of runoff increased. These, in turn, increased sediment loads, caused riverbeds in the Indian plains to aggrade, and led to increases in flooding throughout the Gangetic basin and the growth of islands in the Ganges Delta of Bangladesh.

The 'model' was integrated and the database relatively strong. As Ives and Messerli (1989) stated: "the most compelling and trend-setting characterization of the Himalayan region and its anticipated eco-disaster is that published by Erik Eckholm (1975, 1976)...". Forest cover and flooded areas were being monitored through satellite imagery. Stream gauges were, at least in some locations, in place and had long-term monitoring records. Furthermore, as with the conceptual foundations for the rain-follows-the-plough model, the physical processes were relatively well understood. However, as with that model, interactions with other systems, in this case plate tectonics and the expansion of population into marginal agriculture, were ignored or at least underestimated. As Ives and Messerli pointed out: "The large literature that depicts the imminence of environmental catastrophe in the Himalayan region has tended to confuse cause and effect, has largely missed the essential historical depth, and has assumed the existence of dramatic upstream-downstream interrelationships without requiring rigorous factual substantiation." Subsequent research demonstrated that the causal links between many elements in the model were weak and it was far from clear that forest declines were anywhere near as widespread as portrayed. Furthermore, even if deforestation was causing erosion, natural erosion rates related to tectonic uplift were orders of magnitude higher than the human contribution. Ultimately, the whole basis of understanding Himalayan deforestation came under question.

"The wide uncertainties that currently exist at the bio-physical level - uncertainty as to whether the consumption of fuelwood exceeds or is comfortably within the rate of production, uncertainty as to whether deforestation is a widespread or localized phenomenon, uncertainty as to whether it is population pressures or inappropriate institutional arrangements that lie behind instances of mismanagement of renewable resources ... uncertainty as to whether deforestation in the hills (if it indeed exists) has any serious impact on the flooding in the plains - means that a wide range of mutually contradictory problems are credible." (Thompson *et al.*, 1986).

In sum, the system and the interactions between systems were too complex and poorly understood to be captured adequately. Furthermore, the uncertainty created opportunities for groups in society to advance agendas that matched their worldviews. The international community made major investments in reforestation on the basis of the 'Himalayan deforestation' model. It was practical and pointed to things organizations could 'do' or 'invest in' in order to solve problems. It not only created a problem for organizations to remedy but a problem for them to perpetuate as a means of defining themselves. Ives and Messerli pointed this out: "we must emphasize again that this uncertainty is not merely technical; that it is not just the absence of certainty. Rather, it is structural in the sense that, without their realizing it, certain actors in the Himalayan debate have succeeded in imposing their desired uncertainties within it." Perceptions of environmental degradation became so ingrained in the way organizations approached the Himalayan dynamic that the problems and solutions began to feed on each other. Thompson *et al.* saw it as "generated by institutions for institutions. The survival of an institution rests ultimately upon the credibility it can muster for its idea of how the world is; for its definition of the problem; for its claim that its version of the real is self-evident." The model even became a major factor in arguments by India for the construction of high dams in the Himalayas. These were portrayed as essential to control floods, the fact that the dams would produce large amounts of electricity which India wanted and Nepal could sell was an added benefit.

As the model of Himalayan deforestation has come under question, the interest of international donors in financing reforestation and watershed work has waned. Doubts regarding the 'Eckholm' model led many to question the importance of forestry work in Nepal. However, whether or not they cause flooding in Bangladesh, forestry problems in Nepal are major and have a direct effect on the livelihoods of local populations. Linking these local problems to a regional 'crisis' model may have had short-term benefits where work on forestry was concerned. However, it may have also undermined the long-term focus essential to addressing the real local problems that afflict populations and the forests they depend on in Nepal. This is also a risk in any model posing groundwater overdraft as a major threat to food security.

were noted primarily on the basis of the personal observations of settlers (a situation with many parallels to the current groundwater debate). However, the database was also weak in the case of Himalayan deforestation. Despite massive increases in information gathering technologies, good historical records of forest cover were generally unavailable and data relevant to critical components of the model, such as suspended sediment and baseload transport in rivers, remain inadequate to this date. In many ways, the challenge was one of recognizing that, despite the large amounts of information available, core data relevant to the questions being asked were absent.

Second, both the cases involved the use of explanatory models based on a partial understanding of systems. Although accurate, the physical-process elements underlying the models were partial. As a result, their predictive value was weak. This is also the case with most groundwater systems. Models with a strong predictive capacity are unavailable in all except the most rigorously monitored and analysed aquifers. These aquifers tend to be in wealthy locations such as the Central Valley of southern California, the United States of America, and not in developing countries. Furthermore, even rigorous monitoring and analysis may not enable accurate evaluation of aquifer water availability. For example, in the San Luis Valley of southern Colorado, the United States of America, there is a more than 30-percent gap in water balance estimates despite 40 years of monitoring and analysis driven by litigation over water availability. Experts believe this gap may be related to deep inflow from outside the basin or inaccurate estimates of evapotranspiration from native vegetation, but no one knows for sure. Problems of this type would be exacerbated under conditions in developing countries where groundwater monitoring is a relatively new phenomenon.

Third, as with the case of Himalayan deforestation, answers to the question of the impact of groundwater overabstraction on food security hinge on complex interactions between water-resource, economic and social systems (Plate 6). It is unlikely that all three of these systems are understood to a sufficient degree of precision to develop definitive management responses. Furthermore, even if the systems were understood, interactions between non-linear systems often produce unpredictable and counterintuitive results. This is particularly evident in recent debates regarding the effect of vegetation on stream flows. In South Asia, re-vegetation of watersheds is widely advocated as essential for regenerating springs and river flows. However, studies in Australia document rises in groundwater levels (and the destruction of pastureland through waterlogging) due to removal of tree cover (Moench, 1998). The effects of vegetation on water availability depend on the delicate balance between recharge and evapotranspiration. Improvements in soil characteristics and reductions in runoff associated with vegetative cover generally enhance recharge. At the same time, the vegetation requires water to survive, and evapotranspiration increases. Which dominates depends on a wide variety of factors: species, wind speeds, temperature, soil types, etc. Significantly different outcomes commonly emerge from subtle interactions between such factors in local contexts.

All of the above point to the inherent risks in attempts to link real local problems to global consequences of dubious clarity: it is known that groundwater overabstraction is a major problem in specific regions. Such problems have substantial environmental, economic and other consequences whether or not they have direct implications for global food security. The danger in focusing on macro food-security concerns is that, if these concerns prove open to question, attention will be diverted from groundwater problems that are important in their own right. Furthermore, as in the case of Himalayan deforestation, approaches designed to respond to 'global problems' often obscure responses that could be more effective but would only emerge if the 'problems' were defined in a different way. For example, approaching groundwater overabstraction from the perspective of global food security will tend to focus efforts into

**Plate 6**
**Effluent groundwater seepage maintaining baseflows in Nepal**
*[M. Moench]*

global and national attempts to manage the resource base and control use in ways that maintain local and, by implication, global production. In contrast, if groundwater overabstraction is viewed as more of a regional concern, then approaches that encourage people to adapt to scarcity by migrating or transferring out of agriculture (rather than attempting to maintain production levels) could prove viable.

In order to evaluate whether the above 'cautions' apply to the debate on groundwater overabstraction and food security, Chapter 3 examines the extent to which emerging groundwater problems are understood and the nature of the available data in key regions.

# Chapter 3
# Understanding the dynamics of groundwater resources

The complexity of flow within aquifers may require extensive data and detailed modelling to answer development questions. Even with this, as mentioned above, accurate analysis of the water balance is often complicated by inflows and losses that are difficult to identify, monitor or interpret.

However, relatively simple data, such as specific water levels in a carefully designed network of monitoring wells, can be combined with estimates of rainfall input to provide key indications of groundwater dynamics. Long-term declines in water levels are often indicative of overabstraction conditions. Similarly, stable water levels generally indicate that inflows are in balance with outflows. However, this is not always the case, e.g. where water in confined aquifers is released through compression, water levels may not decline at least initially when overdraft is occurring.

Unless monitoring networks are designed carefully and the data subject to careful analysis, water-level changes can be difficult to interpret. In many cases, declines represent local or regional cones of depression created by the lagged nature of aquifer responses to pumping or changes in inflow, not actual overabstraction. Aquifers can take hundreds of years to equilibrate to changes in extraction and recharge. Similarly, poorly designed monitoring wells often tap multiple aquifers; water levels observed in such wells provide a difficult-to-interpret reading of the pressure in all rather than a clear reading of head changes in any single one. Despite the care required in interpreting water-level changes, wells do provide the best, directly observable data on changes in groundwater conditions. Furthermore, they have a critical advantage in that they relate directly to water access. When water levels fall, whether or not overabstraction is occurring, the cost of pumping increases, shallow wells go dry and groundwater contributions to environmental values (such as the baseflow in streams or wetlands) decline. Furthermore, water-level changes are a key indicator that flow patterns are changing and that low-quality water may be mobilized. For this reason, they can provide advance warning of pollution and quality declines. As a result, water-level changes relate much more directly to 'what people care about' than do more abstract descriptions of extraction and recharge or flow dynamics. Based on these considerations, long-term records of water-level changes are probably the single most important piece of data necessary to monitor changes in groundwater availability. However, having such information is only the first step. In addition, there needs to be a clear mechanism to feed the information into formal planning processes. The mechanism also needs to ensure that data are transparently available to the user groups whose consensus would be required for effective action to limit drawdown to commonly acceptable levels.

## Nature and limitations of available data

At the outset of this project, the hope was to locate significant amounts of data on groundwater levels in the focus countries through publicly available sources, particularly the Web sites and databases of the major international groups such as the Global Water Partnership. It was also hoped to locate substantial additional information through contacts in key organizations, e.g. the Central Ground Water Board (CGWB) in India, the IIMI, the IFPRI, the USGS, the British Geological Survey, the World Bank, and FAO. However, despite substantial effort, this search yielded relatively little primary data. In many cases, estimates of extraction and recharge are difficult to interpret from primary data. In others, there are statements such as: *"water levels in key aquifers are declining at rates of 1-3 m/year"*. However, such broad averages hide substantial variation.

In the 1980s, the UNDTCD (now merged into the DESA) compiled a systematic survey of groundwater occurrences on a country-by-country basis under the Water Series publications. This comprehensive work has not been updated, but at the time it represented the only systematic portrait of groundwater occurrences with an indication (where possible) of trends in groundwater resource use.

The degree to which this broad array of information on specific aquifers and aquifer systems and specific scientific themes related to groundwater can be collated to present a coherent 'state of the art' on the resource base is questionable. The proliferation of relevant material at country level would overwhelm a global exercise. Moreover, the benefits would be negligible compared with the urgent local problems on which groundwater management has to focus.

Box 3 is illustrative of the challenges that data limitations impose on management. Sources contacted at the World Bank, the IFPRI and the IIMI indicate that groundwater data are highly scattered and that access to much of the data is restricted owing to the confidential nature of many consulting reports. Data are often available in state and local-level groundwater

**Box 3: The significance of groundwater data and its uncertainties**

The problems of compiling groundwater data and interpreting abstraction records to establish and model the status of an aquifer system are well illustrated by interim reports on the Northwestern Sahara Aquifer System (SASS/OSS, 2001). This massive system covers Algeria, Tunisia and the Libyan Arab Jamahiriya and broadly comprises two super-imposed sandstones: Continental Terminal and Continental Intercalaire. Up to 8 000 borehole records and associated abstraction data from 1950 to 2000 have been compiled and a regional groundwater model constructed (Modflow - the standard finite difference groundwater model). A preliminary analysis of abstraction data indicates a two to threefold increase in pumped volumes throughout the aquifer system beginning in the late 1970s, peaking in 1990 and thereafter showing stabilization or slight decline. Abstraction from the system as a whole is estimated at 80 $m^3/s$. In 1950, abstraction was estimated at 13 $m^3/s$ and in 1975 had reached 25 $m^3/s$. The impact on the overall water balance of the system is being refined through the application of a regional hydrogeological model. However, the distribution of boreholes indicates that the generation of drawdown externalities will be very localized. At control-point observation boreholes, there has been a marked lowering of the piezometric surfaces since 1980, with total drawdowns since 1950 typically of the order of 20-40 m in the exploited aquifer blocks. However, the control on abstraction data is variable (in Tunisia abstraction yearbooks have been published since 1973). In general, the levels of uncertainty associated with the data derived from a variety of data sources from the period of record are manifold. The authors of the reports emphasize the fundamental methodological differences that need to be appreciated when dealing with hydrogeological as opposed to hydrological data; specifically piezometric (water pressure head or level) altitude corrections and methods of analysis and validation in relation to hydrogeological time series data. For these reasons, the error terms that need to be attached to any hydrogeological observation are significantly higher than those normally associated with standardized streamflow data. Thus, even where great effort and thought is put into standardizing raw hydrogeological data in preparation for modelling activities, levels of uncertainty will remain high. In this particular case, model results do show a good match with the control observation data and thus provide a broad picture of the aquifer's evolution as development has proceeded. However, the data and model results would probably be too coarse for the establishment of quantitative pumping rights in specific aquifer blocks.

departments but access often requires formal government permission and fluency in the local language. Furthermore, in many cases, the data are not in digital format. Hence, accessing data is frequently a time-consuming process.

The primary type of publicly available information is that contained in syntheses such as those produced by regional governments and by Postel (1999), Seckler *et al.* (1999), and Shah *et al.* (2000). The following section on China is typical of the results from these syntheses.

## Groundwater in China

*"Although it is perhaps the most visible manifestation of water scarcity in China, the drying-up of the Yellow River is only one of many such signs. The Huai River, a smaller river situated between the Yellow and Yangtze rivers, was also drained dry in 1997, and failed to reach the sea for 90 days. Satellite photographs show hundreds of lakes disappearing and local streams going dry in recent years, as water tables fall and springs cease to flow. As water tables have fallen, millions of Chinese farmers are finding their wells pumped dry"* (Brown and Halweil, 1998).

The above syntheses of China's data paints a picture of a country facing a serious water and population imbalance. Compounding the problem of falling groundwater levels (Figures 2a and 2b) is the country's huge population and its rapid urbanization as people leave the countryside to seek higher incomes in the industrial sector. This, in concert with population increases (by 2010 China's population is expected to grow by another 126 million), suggests a precarious future for the country's food security. The country's growing affluence compounds the issue as more affluent people consume larger quantities of livestock products (Brown and Halweil, 1998). China's cities are facing severe water shortages as demands for modern amenities grow and as consumption patterns change. Since 1965, the water table under Beijing has fallen by nearly 59 m, dropping 2.5 m in 1999 alone. Competition for water has usually seen farmers lose out to the much more profitable industrial sector. While figures may be imprecise, they do point to the drastic shortage China is facing and the direct implications for grain production. Falling groundwater levels could imply dire consequences for China's food security. China could be forced to import as much as 370 million tonnes of grain per year to feed its population in 2025 (above), with consequent steep increases in cereal prices and disruption of the world market. The country's vast size, growing population, and increasing affluence point to the critical need to make accurate assessments of China's patterns of grain production and water demand. Although the above analysis may be compelling, its accuracy is limited by the available data. National assessments of water demand in China (e.g. IWHR, 1998) have tended to overproject water demand. Moreover, such assessments obscure local aquifer dynamics where points of competition may be intense and shortfalls in supply apparent, but taken together do not necessarily permit a conclusion at macro-level or point to a national structural gap in groundwater availability.

### Problems in accessing groundwater data

The problem of accurate assessment is illustrated in a report by the Ministry of Water Resources (MWR, 2001) on a water sector strategy for north China, specifically the Hai, Huai and Yellow (3-H) river basins. According to this report, the scale of abstraction for urban and rural use in the Hai basin has grown over the past 40 years, but projections of withdrawals in relation to available resources have changed considerably over time (Table 1).

**Figure 2a**
**Difference between shallow groundwater levels in 1958 and 1998 in the Hai Basin Plains in northern China**

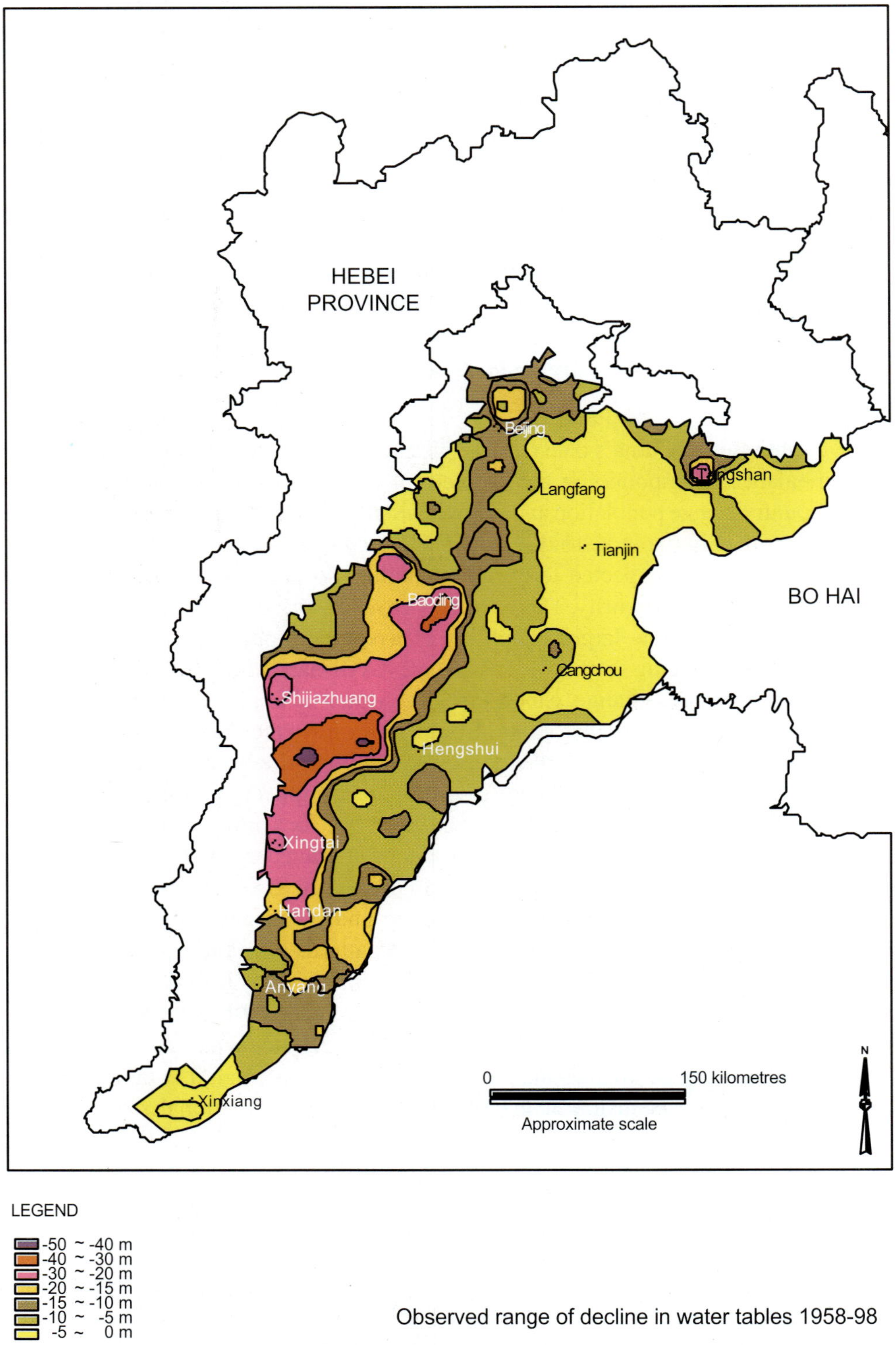


Source: Agenda for Water Sector Strategy for North China. Ministry of Water Resources, World Bank and AUSAID, 2001.

FIGURE 2B

**Difference between deep groundwater levels in 1958 and 1998 in the Hai Basin Plains in northern China**

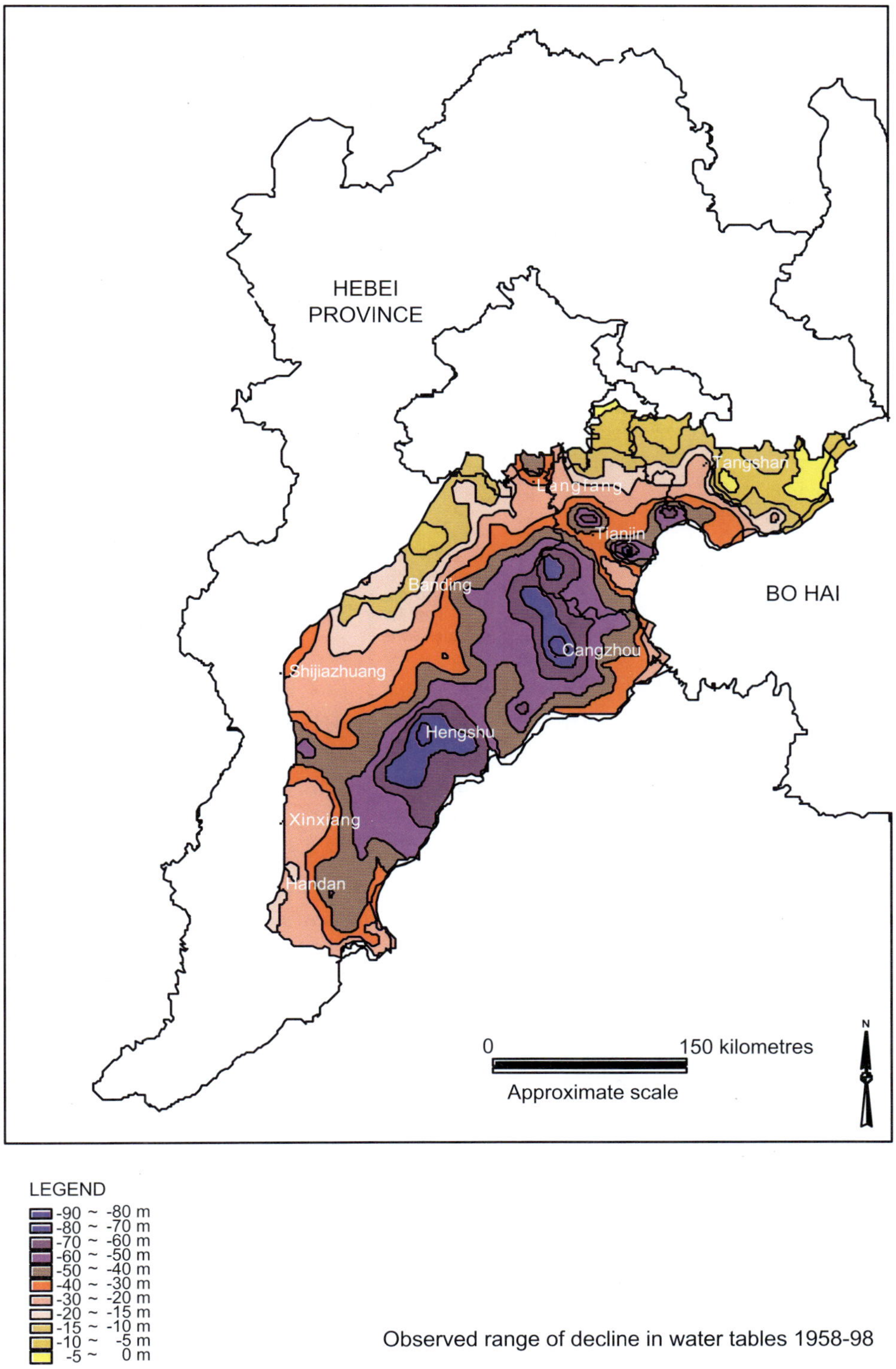


Source: Agenda for Water Sector Strategy for North China. Ministry of Water Resources, World Bank and AUSAID, 2001.

**Table 1**
**Withdrawals for the 3-H basins, 1980, 1998 and projections**

| | 1980 | 1998 | 2000 | 2010 | 2020 | 2050 |
|---|---|---|---|---|---|---|
| | | | (1 000 million $m^3$) | | | |
| **Non-agriculture** | | | | | | |
| IPPDI (1992) | 19.7 | 36.2 | 46.9 | | | |
| UNESCAP (1997) | | | | 45.0 | 75.7 | |
| IWHR, NIHWR (1999) | | | 46.9 | 67.5 | | |
| IWHR (1999) | | | | | 74.2 | 87.6 |
| IWHR action programme studies | | | 39.9 | 54.0 | 68.4 | 88.7 |
| World Bank | | | 37.3 | 43.1 | 52.7 | 59.1 |
| **Agriculture** | | | | | | |
| IPPDI (1992) | 106.4 | 102.5 | 128.7 | | | |
| UNESCAP (1997) | | | | 125.9 | 129.1 | |
| IWHR, NIHWR (1999) | | | 116.5 | 118.3 | | |
| IWHR (1999) | | | | | 109.5 | 106.5 |
| Penman/FAO approach | | | 141.1 | 141.0 | 138.1 | 140.1 |
| **Total** | | | | | | |
| IPPDI (1992) | 126.1 | 138.7 | 175.6 | | | |
| UNESCAP (1997) | | | | 170.9 | 204.8 | |
| IWHR, NIHWR (1999) | | | 163.4 1 | 85.8 | | |
| IWHR (1999) | | | | | 183.7 | 194.1 |

Notes: Sources in table: IPPDI, China's water resources development, 1992. UNESCAP, China: water resources and their use, 1997. NIHWR, IWHR, China water supply and demand in 21st century, 1999. IWHR, Strategic options for the water sector, 1999. IWHR, Industrial and domestic water demand forecasts, project working paper, 1999.
Source: MWR (2001).

The report goes on to emphasize that: *"The two most striking aspects of these results are (a) they all exceed likely future water supply capacity .... by a wide margin, and (b) each successive projection resulted in lower future demands than its predecessors. For example, IPPDI (probably writing in the early 1980s) projected demand to total 175.6 Bcm for the entire 3-H region in 2000. The United Nations Economic and Social Commission for Asia Pacific (UNESCAP), based on analysis done by NIHWR about 1996, reduced this to 170.9 Bcm. IWHR, working on the Action Program in 1999, came up with 163.4 Bcm. The years beyond 2000, where estimated, show similar trends. The Australian Consultant, using an entirely different approach, came up with lower numbers still. ... the 3-H region has probably hit a "brick wall" supply constraint, which in the absence of alternative water supplies, will constrain withdrawals to something near 140 Bcm per year"* (MWR, 2001).

Syntheses such as the above were developed in large part by collecting information from a mix of raw water-production data from provincial authorities and related reports in the consulting, international development and scientific literature. Moving beyond them to tease out the specific groundwater abstraction data would require access to primary sources. This is complex. The same report quoted above states that: *"Effective management (of groundwater) is highly dependent on appropriate reliable and up-to-date information. Currently there are thousands of local and personal databases storing key technical and licensing data in a very unsatisfactory manner. An absolutely fundamental need for effective groundwater management and protection is a comprehensive, publicly accessible, groundwater database (GDB). The complete lack of a GDB is seriously constraining the formulation and implementation of effective groundwater management throughout China. The inability to access information, which at times is part of institutional secrecy, encourages inaction or incorrect decisions. GDBs are well established in almost every country where significant groundwater is used. The lack of such a database in China is surprising."*

Nonetheless, the report does give a summary of groundwater resources and use in the 3-H basins. However, in view of the data problems emphasized in the report, it is not clear whether the summary is any different in its accuracy (or relevance to management in specific locations) from any of the preceding summaries.

## Implications for food security

Even with free access to primary data, substantial questions exist regarding the extent to which a clear understanding of emerging groundwater overabstraction concerns and their implications for economic development or food security would emerge. The MWR (2001) report notes: *"There is hardly a sector in which water does not play an important role, and given its scarcity it may therefore appear very surprising that there is so little correlation between growth and water conditions. But analogy with other poorly endowed countries and regions in East Asia and elsewhere suggests that neither land nor water constraints need be decisive. Rather, it is a question of how well a country or region adapts to its resource endowment that determines whether land and water constraints impact seriously on economic development. ... In this sense, natural resources are no different from the other factors that help determine comparative advantage, and alarmist conclusions to the contrary are very misleading. It is the conclusion of this report that economic growth in the 3-H region ... including agricultural growth, can be sustained and that water may not impede this growth provided immediate attention is directed to finding real solutions to current water pollution and water scarcity problems."*

However, the concluding chapter of the report (on proposed action plans) reports the following contrasting comment: *"The very serious and largely irreversible falling groundwater levels throughout the North China Plain demands a major program of groundwater management planning to reduce groundwater use to sustainable levels. The groundwater management strategy proposed will require a huge effort, however the consequences of not doing it will have major long term implications, such as effectively destroying the groundwater dependent agricultural base, massive subsidence and sea water intrusion, virtual elimination of groundwater as a water source for many cities and countless households and the loss of "insurance" water for future generations."*

The above comments highlight a basic tension in the MWR report that is also present in debates on the relationship between water availability and food security in many other regions. Groundwater overabstraction in the 3-H region is substantial and could lead to reductions in food production. However, even under a severe drought, any reduction in irrigated area would almost certainly occur gradually because water levels are declining at different rates in different locations and water availability varies widely in different parts of the aquifer. This gradual decline could be offset by China's plans to import water from other areas. It could also be offset by a gradual increase in food purchases from global grain markets. Such purchases might in turn encourage more production in other parts of the world and could lead to a broad redistribution of production to areas with greater comparative advantage for irrigated agriculture. At least in the short term, the main threat to food security would occur if groundwater depletion occurred in conjunction with a wide variety of other factors that limited China's ability to adapt. For example, severe food security problems could arise if an extended and severe drought occurred in areas suffering from overdraft and if access to global markets were constrained by war, economic conditions or sudden increases in demand from other countries in addition to China.

Once again, the point is that groundwater overdraft is likely to become a major source of food insecurity primarily if it occurs in conjunction with other factors that limit China's ability to adapt.

### The viability of groundwater management

In addition to uncertainty over the implications of groundwater overabstraction for food security, the way towards management 'solutions' is not clear. For example, the MWR (2001) report advocates a set of regulatory measures predicated upon identification and planning of sustainable yields and the coordinated compliance of millions of groundwater users. As it states: "*The fundamental objective proposed is to reduce groundwater use to sustainable levels by 2015.*" A broad range of technical, institutional and management actions are required to achieve this goal.

"*The key actions required are:*

- *All significant groundwater usage areas be defined as Groundwater Management Units and the Sustainable Yield be determined.*
- *Groundwater management plans be prepared and implemented as per the program described herein.*
- *Allocation licensing be linked with the sustainable yield assessment.*
- *Allocation licensing be only undertaken by one department.*
- *Licensing of well construction drillers.*
- *A National Groundwater Data Base be developed.*
- *A groundwater pollution prevention strategy be prepared.*
- *The Ministry of Water Resources should review the adequacy of regulations required to implement the groundwater management reforms proposed herein.*
- *The introduction of realistic groundwater prices.*
- *A major education program about groundwater processes and the need for groundwater management.*"

Despite the above recommendations, the report does not address in detail the viability of the management actions it advocates. However, it does hint at the probable complexities when it states in the case of shallow aquifers that: "*In principle, shallow wells should be regulated and charged for in a comparable manner to deep wells, but in practice regulation of dispersed small wells may be difficult and farmers may resist resource charges. Unsustainable use results in well interference, higher pumping costs, reductions in over-year storage and other adverse externality effects (of which mobilization of arsenic and other minerals may be the most damaging). Nevertheless exploitation of shallow aquifers tends ultimately to be self-regulating as farmers adjust conjunctive use of rainfall, surface supplies and groundwater to whatever combination of resources is available to them. This may not result in a theoretical "optimum" but in practice can be an acceptable compromise.*"

More indications of the probable challenges are contained in a recent study of the Hauihe basin (United Nations, 1999). The report emphasizes that, even with centuries of experience in hydraulic management and the proliferation of voluminous hydraulic regulation, surface water management in the Hauihe basin is effectively anarchic and that implementation of sensible water policies and law remains blocked by the economic self-interest of riparian provinces.

Given the fragmented database in the 3-H region, the technical viability of some key recommendations (such as assessment of sustainable yields) is unclear. As the situation in the Hauihe basin illustrates, the social and political viability of most of the other core recommendations may also be open to question. Where surface-water management is, for a combination of social and political reasons, effectively anarchic, there is little reason to expect that management will be more effective in the groundwater case. Hints of the political complexity of groundwater management are also evident in the contradictory objectives evident at numerous places in the report. For example, in relation to the tension over groundwater allocation to different sectors, the MWR report states: "*From a macroeconomic view, the vastly higher economic values of water generated by the urban and industrial sectors should not be sacrificed for increased agricultural production. The ability of China's industrial sector to generate trade surpluses easily offsets foreseeable reductions in agricultural output that may arise from irrigation water shortages. However, three-fourths of China's population is rural and depend on agriculture for two-thirds of their incomes. Protecting and increasing these incomes and maintaining growth in agricultural production is a national concern, and one that depends in part on more and better irrigation.*"

Overall, the emergence of major overabstraction problems contrasted with uncertainties regarding their real extent, the issue of what can be done (in terms of implementing regulation) and the implications of leaving problems unaddressed generates substantial tension in debates on approaches to groundwater.

The circumstances in northern China illustrate the complexity inherent in any evaluation of the implications of groundwater overabstraction for regional food security. Questions start with basic information and data on groundwater conditions and then move upwards through the implications of water availability for agriculture and economic development before approaching the food security equation. Questions also depend on the viability of management. A key starting point in investigating this chain rests on groundwater data, on the degree to which emerging groundwater problems are understood and their implications open to interpretation.

## Groundwater monitoring in India

In India, the groundwater data issue has been examined in detail and there exists a national organization dedicated to the collection and analysis of groundwater data. The case of India is illustrative of the inherent data problems facing any global evaluation of groundwater conditions.

India maintains extensive networks of wells for monitoring groundwater levels and water quality. The CGWB operates a network of about 10 000 monitoring wells nationwide. Groundwater entities attached to various state-level departments operate other networks. Water is a state matter under India's constitution but the central government provides much of the financing for groundwater development. In consequence, there is substantial competition for control over funding and the interpretation of groundwater data. This situation complicates scientific evaluation of groundwater conditions.

Historically, most of CGWB-maintained wells were open dug wells (generally used for drinking-water). However, a substantial number of isolated piezometers have been installed over the past five years and the number is increasing. Aside from a limited selection of piezometers and key wells that are monitored more frequently, most wells in the CGWB network are monitored four times a year, including once before and once after the monsoon.

Groundwater data collection and monitoring by organizations at state level is relatively similar in design to the CGWB network but varies substantially in relation to the attention given to it. Some states have developed extensive groundwater departments attached to major line organizations such as the Public Health Engineering Department. Other states maintain small units as adjuncts to marginally related mining or geology organizations. Most states monitor a network of 1 500-2 000 wells although some have many more (Rajasthan monitors more than 6 000 wells). Groundwater levels in state networks are generally monitored twice a year, before and after the monsoon. As with the CGWB, state monitoring networks are dominated by open dug wells supplemented by a small number of piezometers. Some are moving to strengthen their networks through the installation of piezometers (Rajasthan being a prime example). However, most of these piezometer networks are only a few years old.

Until recently, data collection by both the state organizations and the CGWB was organized on the basis of blocks, local-level administrative units between the *panchayat* (local government) level and the major districts into which each state is subdivided. Recently, some states have transferred the focus of groundwater data collection and monitoring to hydrological units. For example, Maharashtra conducts monitoring on a watershed basis, while Rajasthan uses 'hydrologic potential' units. Recent revisions in groundwater evaluation methodologies by the Ground Water Resource Estimation Committee (GWREC) recognize the importance of hydrological units and recommend that monitoring and evaluation be based on watersheds (GWREC, 1997). Regardless of the base unit, groundwater conditions in most states are reported on the basis of administrative blocks. This reflects the core reason why groundwater monitoring was initiated: its role in guiding the allocation of groundwater development financing.

**Analytical methods and the role of groundwater data in development finance**

Groundwater development in India was initiated on a large scale in the 1950s and 1960s as a core strategy to address famine and increase agricultural production. The spread of groundwater irrigation was achieved primarily through government credit and subsidies to farmers for wells, through extension of the electricity network into rural areas and through power subsidies for both diesel and electricity. Subsidies and credit guarantees for new wells are provided through the National Bank for Agriculture and Rural Development. These are targeted to local banks at the block level in rural areas. In order to target available financing as efficiently as possible and avoid financing where groundwater resources are limited, funds are allocated on the basis of groundwater availability estimates.

Availability is defined as the balance between recharge and extraction. Estimates of this balance are defined as the 'level of development', equal to the 'net yearly draft' divided by the 'utilizable resource for irrigation' (GWREC, 1997). The GWREC categorizes areas for the purpose of financing as 'safe' (level of development less than 70 percent with no significant long-term decline in water levels), 'semi-critical' (level of development between 70 and 90 percent and with no significant long-term decline in water levels) and 'critical' (level of development between 90 and 100 percent with declines in water levels, or more than 100 percent and no declines in water level). Areas where the level of development exceeds 100 percent and both pre- and post-monsoon water levels show long-term declines are classified as overexploited.

In general, financing for groundwater development is open in areas where development estimates are below 70 percent, somewhat restricted in critical areas, and highly restricted in overexploited areas. Local surveys are required in critical areas before the financing of any further groundwater development. For overexploited areas, the GWREC (1997) recommends

**PLATE 7**
**A groundwater recharge message in Gujurat, India**
*[M. Moench]*

that "*there should be intensive monitoring and evaluation and future ground water development be linked with water conservation measures*" (Plate 7).

## Recharge estimation

The estimated level of development plays a central role in the allocation of funding. The recharge component of this is estimated using the 'water-table fluctuation method'. The core principle is that water-level changes between the pre- and post-monsoon readings (generally a rise) are multiplied by the estimated specific yield of the regional formation and by the area of the evaluation unit (watershed or block) in order to estimate the volume of water recharged during the monsoon. This value is normalized to reflect the nature of the rainfall year, and corrections are made to account for recharge during the non-monsoon period and other inflows. Finally, a percentage of the estimated water is reserved for drinking. The remainder is classified as the net resource available for irrigation. Extraction estimates are based on well surveys and estimated pumping hours (GWREC, 1997). In most states, the above method has been applied to estimate groundwater availability in administrative blocks regardless of the nature of the hydrological units involved. Water-level fluctuations in the 5-10 monitoring wells in each block, together with rainfall (used in normalizing fluctuation measurements), represent the only direct measurements. The remainder of the estimate depends on assumptions derived from other studies regarding the specific yields, pumping hours and other factors applicable to the area.

Estimates of gross recharge based on water-table fluctuations contain large scientific uncertainties. First, the water-table fluctuation approach assumes that resource availability can be calculated on the basis of changes in saturated storage. Particularly in hard-rock areas, changes in storage in the vadose (unsaturated) zone may be the primary factor determining actual water availability (Narasimhan, 1990). Second, estimating changes in saturated storage requires accurate assessments of specific yields from pumping tests. Most analytical methods for interpreting pumping tests were devised for borewells in alluvial aquifers with relatively simple geometric configurations. They do not apply to the large-diameter wells and complex, heterogeneous, hydrological conditions typical of the hard-rock aquifers extending throughout most of peninsular India. As Narasimhan (1990) states: "indiscriminate fitting of hydraulic test data to available mathematical solutions will but yield pseudo hydraulic parameters that are physically meaningless." Therefore, one of the key parameters necessary for estimating recharge using water-table fluctuations may be impossible to calculate in many areas.

### Abstraction estimation

Abstraction estimates represent the 'second half' of the equation for estimating the level of development. Moench (1994) argues that these probably contain more inherent uncertainties than do the recharge estimates. In India, groundwater abstraction is estimated using a combination of well census figures, average well commands, crop areas, water duties, well yields, and pumping hours. The GWREC (1997) recommended calculating abstraction by multiplying the average area irrigated by each well by the average annual irrigation depth. In practice, the precise method used to calculate extraction varies in different states and localities.

Regardless of method, there is substantial uncertainty in the basic data underlying extraction estimates. In Gujarat, the 1986-87 census of minor irrigation was conducted in a drought year when many wells were dry. It is the only comprehensive source on well numbers. If one takes the number of wells counted as functioning in 1987 as the basis for estimating extraction (as was done in the 1992 assessment), then the state's net area irrigated from groundwater is about 1.7 million ha. However, the total number of wells present is much higher than the number functioning during the 1987 drought. If all the wells were included when estimating irrigated area, then the total would be about 3.5 million ha. As the official 1992 estimate of the total area that can potentially be irrigated from groundwater is 2.9 million ha, the difference between the two well counts is a difference between 59 and 121 percent development. Similar problems are also present with crop water duties. These are generally estimated using data from experimental farms. The number of irrigations used on experimental farms and in farmer's fields can vary by a factor of three or more (Goldman, 1988).

Given problems in the available data, most groundwater experts working in government departments consider abstraction estimates to be less reliable than recharge estimates. Accurate estimation of each of the components used for calculating draft would require substantial surveys for which the groundwater departments lack sufficient resources. As a result, hydrologists working for state and central organizations often state that they adjust abstraction figures to represent what they think is happening.

### Methodological improvements

The 1997 GWREC report contains several important methodological improvements over the water-level fluctuation methodology adopted in 1984. The new methodology recommends that evaluations be done on the basis of watersheds rather than administrative blocks and incorporates consideration of long-term water-level trends. The earlier approach did not utilize water-level trends in the definition of critical and overexploited areas. Instead, it focused on the estimated 'level of development' within administrative blocks as the sole measure.

However, major methodological issues remain. All of the inherent data issues in estimating recharge using water-level fluctuations remain. Groundwater abstraction volumes are also highly uncertain. The report states that abstraction is not to be estimated "based on (a) electric power consumption from pumpsets, (b) statistics of area irrigated by ground water and the associated crop water requirements, and (c) use of remote sensing data to obtain seasonal data on area of different irrigated crops in non-command areas, where only ground water irrigation is used." The report goes on to indicate that: "In view of the uncertainties in the estimation of ground water draft by any of these methods, it is clearly desirable to use more than one method for draft estimation to enable a cross check." As power for groundwater pumping is unmetered in most areas and many wells are diesel, the first method is of limited usefulness at present.

Statistics on crop water requirements and irrigated areas have the same inherent issues mentioned by Moench (1994). The use of remote sensing information might improve area estimates but would not improve crop water use estimates.

Beyond the methodological issues, many states are raising questions with regard to implementation of the recommended methodologies. Such questions have made it impossible for the CGWB to issue any updated national evaluation of groundwater resource conditions.

The only relatively recent and publicly available report (CGWB, 1995) is based on data from as early as 1989-1990 and was compiled using the older estimation methodology that neglected long-term water-level trends.

**Data collection and quality**

The above methodology and its utilization as a key tool have been the major factor driving groundwater data collection in the past 20 years. Because 'level of development' was the main criteria for allocating groundwater development financing, the focus of groundwater monitoring was to collect the water-level fluctuation data necessary to drive the methodology. Furthermore, arid states such as Rajasthan and Gujarat, which depend heavily on groundwater, had a strong incentive to generate their own data as a tool in the debates with the central government over development finance. As a result of this dynamic, pre- and post-monsoon water levels are available as the core data set with state groundwater organizations and the CGWB.

Although some wells have been monitored since the mid-1970s, most of the monitoring began in the 1980s after the GWREC report. Staff from the CGWB and state groundwater organizations take measurements manually. Most of the wells selected are dug wells because tubewells are often blocked by pumps. Where possible, wells used for drinking-water rather than irrigation have been selected. This selection rests on the assumptions that drinking uses will result in less extraction and that the well will provide a more accurate measure of water-table conditions than more heavily utilized wells. However, many state networks contain a significant number of irrigation wells where use levels may be high. This is also true of the CGWB network (although this network contains a significant number of piezometers).

Data from the national and state networks are currently being compiled electronically and as part of the Hydrology Project, which is supported by the World Bank. This project covers most states. Although the data are in electronic format, they are not readily available to the public. There are also substantial questions regarding the reliability of data, particularly those from some of the state networks (Moench, 1994). In addition to the usual problems with any form of data collection, the role that data play in determining financial allocations for groundwater development creates a major structural dynamic that can affect their quality. For example, in Gujarat, water levels recorded in the original logbooks were corrected as many as four times from the original measurement. The then chief hydrologist of the state explained this as corrections for staff errors explaining that the water level cannot fall during the monsoon, hence data that indicate otherwise must be errors. However, under the water-table fluctuation methodology, water-level declines during the monsoon would reduce the estimated recharge and could shift the status of a block into the restricted critical or overexploited categories. As a result, there are incentives to manipulate the outcome of groundwater availability estimates and these appear to have affected the quality of basic data at least in the above case. Moench (1994) discusses this issue in detail. However, the main point is that there are significant non-technical factors that may have affected the quality of some of the core water-level data.

**Variability**

Data quality issues are compounded by the large seasonal and interannual variability in water-level trends and hydrological conditions encountered within physiographic regions. Administrative blocks often contain a wide variety of geological formations and hydrological units. In many cases, monitoring wells have not been selected to represent these units but to ensure a reasonable distribution of monitoring points within a given administrative block. As a result, it is difficult to relate monitoring data to specific aquifers or hydrological units.

The lack of good hydrogeological information may be one factor explaining the high degree of variability in water-level trends observed in monitoring wells within relatively small areas. For example, water-level trends within regions often display substantially different hydrographs. For example, maps of ten-year changes in water levels made by the Ground Water Department in Rajasthan indicate broad declines of from 1 m to more than 10 m between 1984 and 1987 (Rathore and Mathur, 1999). A review of the actual data on which this map was prepared indicates that field conditions are more complex. For example, in Jaipur, one of the districts most affected by overabstraction, monitoring wells show highly varied trends and both the rate and extent of decline depend on the time-period selected. Water-level trends are rising in some blocks, while the extent to which water levels are falling is less clear in others. This probably does not invalidate the general picture indicated in the Ground Water Department's map, but it does highlight the large variation present at local levels even within areas where overabstraction levels are reported to be high. This variation complicates estimation of the impacts of overabstraction on water access and thus on food production and security.

In addition to variation in groundwater conditions, there is substantial variability in precipitation, the main source of recharge. This is particularly true in arid areas. As Pisharote (1992) notes for the district of Kutch in Gujarat, half the annual rainfall typically occurs in a period of 2-3 hours in the monsoon season. There are generally 8-10 rainy days in the year and rain actually falls for an annual average of 12-15 hours. Under these conditions, rainfall is highly located and runoff is intense and lasts for brief periods. Rainfall also varies significantly from year to year and location to location. As a result, long periods of record would be required to develop accurate estimates of rainfall, runoff and recharge. This is not a minor concern. As discussed in Box 1, rainfall increases in the High Plains occurred over nearly 20 years and provided an ultimately unfounded proof for many regarding that model's validity. Reisner (1986) reports a similar situation where water of the Colorado River, the United States of America, was allocated between states based on an average annual flow estimate of 17.5 million acre-feet (21 577 500 000 $m^3$). However, the period of record happened to include some of the wettest years in the Colorado basin's history. The water allocation agreement did settle disputes temporarily but, as subsequent records indicate, only temporarily as "the average annual flow of the Colorado River was nowhere near 17.5 million acre-feet." (Reisner, 1986).

The potential for problems similar to those above is high in the case of groundwater in India. Groundwater monitoring data in India span a few decades at most. Given the wide range of hydrological units involved, the high variability of the hydrological cycle in arid areas and the short periods of record, such data must be interpreted with caution. This is particularly true with 'level of development' estimates as they are derived by combining a wide variety of information, all of which is subject to substantial regional, seasonal and long-term variation. Furthermore, extraction and recharge estimates do not reflect directly the factor that has the greatest influence on access to water, i.e. water levels. Users may not care about the theoretical balance between extraction and recharge but they do care about water being available in the wells, and this availability is a direct function of water-level fluctuations, not the longer-term balance between extraction and recharge.

### Water-level fluctuations

Hydrographs for locations in Gujarat illustrate some of the water-level conditions encountered through monitoring open dug wells in hard-rock areas. In Gordhanpur, monitoring between August 1981 and August 2000 indicated a water-level decline of about 2 m in a 20-year period (Figure 3). However, the main information this hydrograph conveys concerns daily fluctuation rather than long-term change in average levels. During dry periods, e.g. 1986-88, water levels often fell more than 20 m below ground level. However, they approached the ground surface at other times. The water-level observations reflect a rapid aquifer response to changes in conditions, presumably rainfall and extraction. The declining trend line may be an artefact of the period of data collection rather than an actual trend. This may also be the case for the Jhalod hydrograph (Figure 4), where the trend line is increasing and fluctuations are significant.

The trend in the Junagadh hydrograph (Figure 5) indicates a decline of about 3 m in the period 1970-2000. However, the hydrograph indicates that water levels return to within 3-7 m of the ground surface in most years. In 1975-76 and in 1986-88, water levels fell to more than 20 m below ground level. This suggests that the aquifer was responding rapidly to drought or short-term increases in extraction. The hydrograph also suggests that the magnitude of annual fluctuations in water level has increased in the past decade, but the data are not conclusive. However, such a pattern would be consistent with that expected in hard-rock areas as development increases. Well discharges typically increase linearly with increases in the width and number of fractures and, reflecting declines in fracture width and spacing, decrease linearly with increases in depth to the static water level (Basak *et al.*, 1993). Therefore, the magnitude of fluctuations is likely to increase as well depth increases, regardless of whether or not recharge increases.

The hydrographs in Figures 3, 4 and 5 are all from open dug wells where use may be a factor in the water levels observed on any given day. Continuous monitoring in isolated

**FIGURE 3**
**Hydrograph of Gordhanpur (41J3A02)**

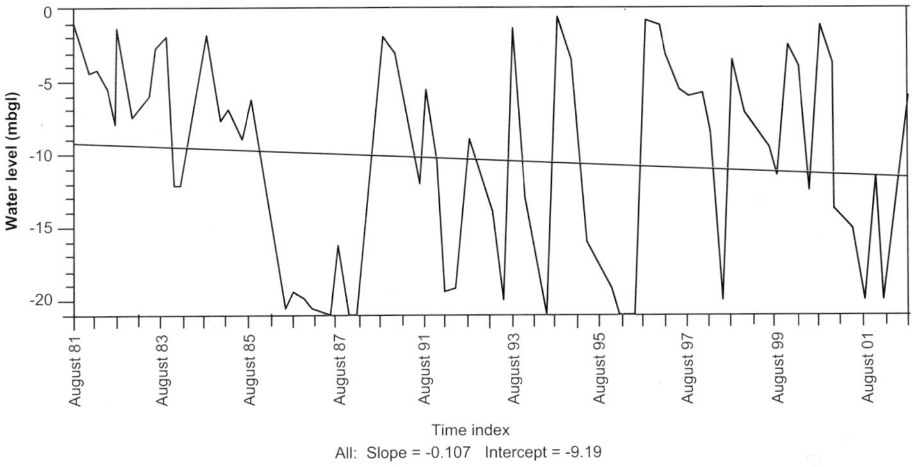

**FIGURE 4**
**Hydrograph of Jhalod (4614A01)**

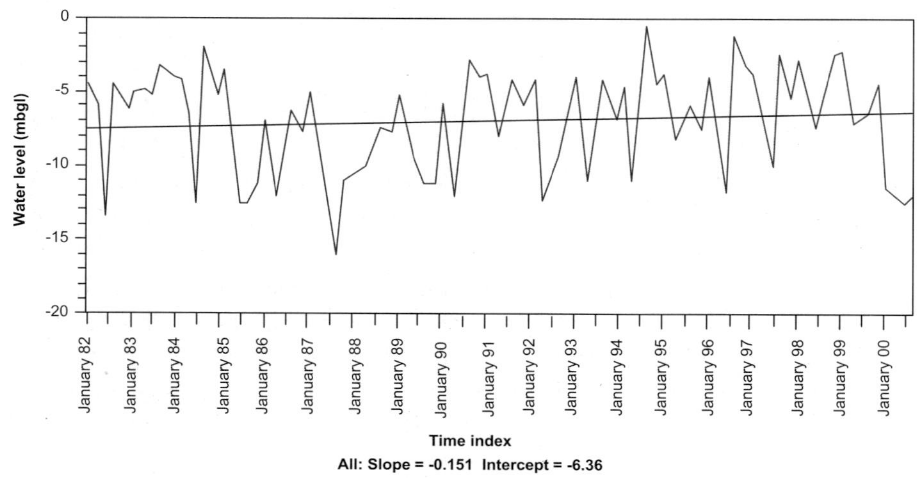


**FIGURE 5**
**Hydrograph of Junagadh**

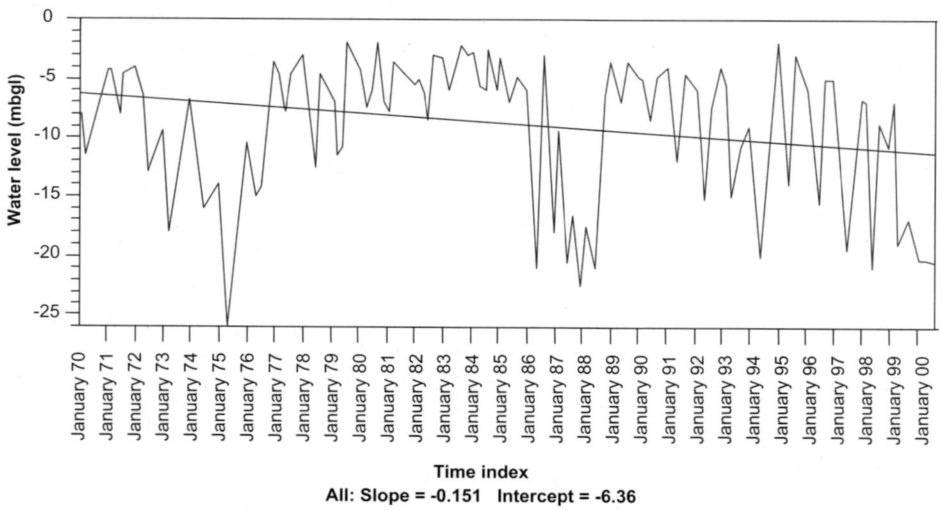


piezometers typically shows much less fluctuation and might display the long-term water-level trends observed in adjacent dug wells. This suggests that the water-table fluctuation method contains inherent sources of error in hard-rock areas. Because permeabilities in hard rocks are often very low (dependent on whether or not wells intersect major fracture zones), water-level fluctuations in dug wells could indicate little concerning actual changes in regional storage. Instead, they would be dominated by recent use, and fluctuations would reflect changes in storage within the well itself (as with a cistern) rather than in the surrounding area as a whole.

Speculation of the above type is difficult to resolve within the limitations of available data. As water levels are monitored only four times a year (and often only twice a year), data documenting the fluctuations necessary to relate water-level changes to extraction and recharge events are lacking. However, it is important to recognize that fluctuations of the type shown in Figures 2, 3 and 4 have much more significant implications for the actual water availability experienced by farmers. Under these circumstances, the value of annual or even interannual aquifer water balances is limited.

Dug wells are often only 10-20 m deep. If fluctuations such as those observed in Junagadh, Jhalod and Gordhanpur are common, wells will frequently go dry irrespective of whether estimates indicate the presence of overabstraction or whether long-term water-level changes show declines. In many areas, farmers complain about water-level declines and the drying out of wells, but estimates of the 'level of development' (and long-term water-level trends) indicate little problem. This may be the primary factor underlying conflicting views of scarcity and availability. In many cases, water-level fluctuations, including seasonal ones, may have more significant implications for water availability than lumped estimates of recharge and extraction.

## Implications

Despite recent methodological improvements, estimates of the 'level of development' (i.e. extraction and recharge estimates) almost certainly remain unreliable. In the case of recharge estimates, this is due to their dependence on aquifer parameters, such as specific yield (which varies considerably between locations), data quality concerns and uncertainties in what was being measured (actual changes in water levels or localized drawdown). Major uncertainties are also inherent in extraction estimates because these depend on limited surveys and the results of experimental farm research (often significantly different from actual use by farmers). Moench (1991; 1994) discusses these and other uncertainties inherent in the water-level fluctuation approach in detail.

Although the new methodology outlined in the 1997 GWREC report incorporates evaluation of water-level trends, it does not eliminate many uncertainties. Data quality concerns remain major as does the short period of record for most of the monitoring network. Furthermore, in hard-rock areas, water-level fluctuations may be as or more important in relation to the actual amount of water available in wells at any given time than longer-term water-level trends. Water scarcity may be severe during droughts or on a seasonal basis even where long-term trends are absent and water-balance estimates indicate substantial resources are available for development.

Resolving the above uncertainties would require long-term data on water levels that capture seasonal fluctuations. Such data are currently unavailable and could take decades to collect. Despite their limitations, water-level data in the databases maintained by state groundwater organizations and by the CGWB would probably be sufficient to give a broad picture of water-level changes over recent decades. However, accessing this data would require government permission to utilize the database developed by the Hydrology Project. It would also require visits to state groundwater organizations not involved in the Hydrology Project. In addition, as the process of data entry from states is ongoing, visits to state groundwater organizations and CGWB offices in those states involved in the Hydrology Project would also be required in order to confirm that all relevant data are incorporated in the database. Finally, even with sufficient resources to visit all the relevant departments, there may remain extensive clearance processes that impede access to the data.

## Evaluation summary

The following sections summarize the key issues relating to the evaluation of groundwater conditions in India.

### *Data*

The available groundwater database contains a wide array of inherent limitations including:

- Short period of record with regard to water-level trends.
- Limited monitoring within the year eliminates the ability to analyse the magnitude and extent of seasonal water-level fluctuations.
- Technical questions regarding data quality (e.g. monitoring in dug wells that may be subject to heavy use).
- Non-technical factors affecting data quality (e.g. pressures to modify data to influence the allocation of funds).
- Limited information on the hydrogeological context and formations in which data are collected.
- Insufficient monitoring to capture the regional and seasonal variation in groundwater throughout India.

### *Analysis*

The water-table fluctuation method used to evaluate extraction and recharge has inherent limitations. While such a water balance approach represents a logical, physically based construct, many parameters are subject to substantial (and generally unevaluated) uncertainty, and key processes (such as lateral inflows and outflows) are not captured. Equally, it is good practice to test a range of methods for testing water balances. Experience in other locations indicates that substantial uncertainty often remains in water-balance estimates. This is the case with the San Luis Valley of southern Colorado, the United States of America, where decades of detailed monitoring have been combined with the application of physically based hydrogeological models constructed as a result of intense legal debates over water management. Therefore, the accuracy of water-balance estimates generated using the water-table fluctuation method is almost certainly poor. Computation of confidence intervals for extraction-recharge estimates would provide a useful insight into the statistical accuracy of estimates.

### *Relationship to water availability*

Water-balance estimates often provide little insight into whether or not water users in a given region already face or are likely to face actual problems of water scarcity. However, in many cases the only official data on groundwater availability at a national level is assessed through water-balance estimates.

### *Data access*

Available basic data on water levels and water-level fluctuations are generally inaccessible. Access to core water-level monitoring data compiled by the Hydrology Project and in state government departments is complicated and requires a substantial process of approvals.

### *Politics*

Groundwater information plays an important role in development financing and other politically sensitive decisions. The data are politically sensitive and analyses are subject to pressure (Moench, 1994). As a result, objective evaluation of the extent of groundwater depletion is problematic.

## CONCLUSIONS ON GROUNDWATER DATA AND ANALYSIS

The issues identified above highlight the substantial uncertainties inherent in estimating the areas affected by water-level declines and overabstraction. These uncertainties apply to situations in developing countries and in many industrialized countries.

Most data contained in general reports are highly aggregated and processed. As in the case of the estimates produced by the CGWB and state governments in India, such processing generally involves a wide array of assumptions. It is also frequently dependent on other types of information (such as extraction estimates) where results depend heavily on assumptions and estimation techniques. Periods of record for groundwater data are generally short, and data-quality concerns are often present. In combination, such factors make water-level change and overabstraction assessments based on publicly available information problematic. This is not just a concern for groundwater but is a common feature in many water-resource evaluations. As one author comments: "*The data on the water balance and productivity of water for irrigation systems at basin, system and farm level are scarce... wherever such data are reported the method of derivation is not described....The inadequacy of data make it difficult to analyse productivity at system and basin level*" (FAO, 2002).

Available groundwater information indicates that water-level declines are occurring in key agricultural areas and that some aquifers are almost certainly experiencing high levels of overabstraction. However, the extent of areas where water levels are declining over the long-term is not possible to determine from the information accessed through this review.

Apart from data, the largest source of uncertainty relates to the assumptions underlying the interpretations of groundwater conditions. Distributed-hydrogeological and water-balance models can be a powerful tool for interpreting water-availability issues provided that the storage and abstraction volumes are known and understood with some degree of scientific confidence. As the examples in Boxes 1 and 2 illustrate, models can generate misleading results where key hydrodynamic processes are unknown or misinterpreted. Without detailed hydrogeological studies, it is generally impossible to determine key parameters such as permeability, storage and transmissivity. Equally, evapotranspiration from native vegetation or shallow groundwater tables is rarely evaluated but can represent a major portion of the water balance. Even 'known' portions of the water balance (e.g. crop water use) are often subject to substantial uncertainty because of the difference between conditions on experimental farms and field conditions. Crude water-balance estimates developed in the absence of detailed hydrogeological information may well contain errors as major as those in the models of Himalayan deforestation or agriculturally induced climate change discussed above.

The above comments are not intended to denigrate modelling efforts. Detailed hydrogeological models and water-balance estimates are useful tools for organizing understanding of physical data and interpreting the impacts of use or management changes. When properly developed (calibrated and validated) with suitably precise data, they can also become powerful predictive tools. However, data requirements and the fundamental parameters (permeability, storage, transmissivity and leakage) needed to structure hydrogeological models and define confidence limits are lacking in many analyses of groundwater availability, particularly in developing countries. As a result, regional extraction-recharge estimates provide little insight into actual groundwater conditions. In the absence of long-term data relating water-level changes to specific aquifers, it is impossible to quantify the impact that current patterns of groundwater extraction are likely to have on water availability in key agricultural areas. Groundwater scarcity is emerging in some areas, but the size of those areas and the severity of the threat to irrigation water supplies remain unclear.

# Chapter 4
# Implications

## Implications for the analysis of food security

The groundwater data and analytical issues highlighted in the preceding chapters place major limitations on the analysis of relationships between water and food security. To improve estimates of groundwater availability for irrigation beyond the calculations of Postel (1999), Seckler *et al.* (1999) and Shah *et al.* (2000) would require a major initiative to collect primary groundwater data and the associated information essential to interpret it correctly from widely dispersed locations. In addition to the relatively straightforward process of locating data sources and documents, this would require substantial effort to obtain approval from governments to access primary data. The research process undertaken to produce this paper investigated a wide array of material relating to the links between groundwater access, yields and crop production. However, little of this second-level material has been incorporated into this paper because of the core uncertainties relating to the status of groundwater resources.

Compilation of available case-study data on groundwater from country sources would improve substantially the understanding of groundwater-level trends in key agricultural areas. This would enable evaluation of probable changes in the economics of groundwater extraction and changes in access to groundwater for poor and marginal groups or those dependent on specific technologies such as dug wells. This type of information would be useful for evaluating probable distributional and economic impacts as development proceeds and water levels decline. However, it would not resolve the inherent problems of data quality and short periods of record that are often encountered. Similarly, it would not resolve questions relating to other key components of the water-balance equation, e.g. extraction rates, leakage between aquifers, and evapotranspiration by native vegetation and crops. As a result, while the data should enable improvements in overabstraction estimates, they would not resolve many of the major modelling issues and would probably not enable accurate estimates of groundwater overabstraction at a global or regional level. These types of uncertainties would be magnified if taken a step further and used as inputs for analysis of global food production. The parallels to debates over climate change are worth noting here. According to Rosenzweig and Hillel (1995): "*The uncertainty inherent in predictions is a very important feature of climate change impact studies...Other uncertainties derive from the fast pace and unpredictable directions of future social, economic, political and technical changes. The world of the coming century will be different in many ways; unforeseeable developments in other sectors may change the way in which agriculture responds to climate change. ... An even more challenging task is to estimate the probability of coincidental events that might happen in conjunction with global warming, spanning the range between low probability catastrophic events (called "surprises") and higher probability gradual changes in climate and associated environmental effects. A seemingly small change in one variable – for example, rainfall – may trigger a major unsuspected change in another; for example droughts or floods might possibly disrupt the transport of grain on rivers. Moreover, one "surprise" may then lead to another in a cascade, since biophysical and social systems are interconnected.*"

Given the status of groundwater information at national level and the inherent unknowns in the models for predicting the impacts of development, the above comments also apply to estimates of the impacts of groundwater-level change on global food production and security. Improvements in access to primary data on groundwater would improve models of food production and food security. However, the predictive value of such models will remain limited by data quality issues, incomplete understanding of systems and ongoing processes of climate, demographic, economic and agricultural change. Thus, the usefulness of this type of analysis is uncertain. In the climate change case, Rosenzweig and Hillel (1995) advocate courses of action that respond to this uncertainty and increase resilience: "*Identifying potential surprises and communicating them to the public and policy makers may help to build the resilience that is needed to anticipate and mitigate harmful effects in a timely fashion.*" Similar courses of action appear appropriate in the debate on groundwater overabstraction and food security.

The above discussion suggests that, rather than attempting to analyse the macro implications of water availability or, more specifically, groundwater problems for food production or access on an aggregate level, it would be more productive to focus on a broad array of early-warning indicators that can trigger responses to food security concerns as they emerge at specific points of time in specific local contexts. Food security problems emerge because of a confluence of hydrological, climate, economic and social factors. Therefore, analysis could focus on developing indices of food security vulnerability that combine an array of long- and short-term physical, economic and social indicators. Groundwater conditions and availability would be among the more important water availability indicators that would need to go into this analysis. However, they would need to be combined with other indicators that reflect, for example, drought probabilities, general economic conditions (availability of alternative sources of work), global food availability and transport capacity. Such indices could be used to trigger proactive responses to emerging food security problems before they reach a critical level and thus reduce the need for *post facto* relief programmes. Thus, the role of analysis would move away from efforts to predict quantitatively the impact of groundwater depletion on aggregate food production and would focus instead on the development of more localized early-warning indicators.

Placing greater emphasis on indicators of food security vulnerability does not reduce the importance of groundwater management. While data limitations and other factors restrict the ability to quantify with any degree of confidence whether or not groundwater overabstraction and falling water levels have major implications for aggregate food production and access, it is known that they could. It is also known that water-level changes have major implications for poverty, environmental values, health and regional economies irrespective of whether or not global food security is at risk. Thus, the critical importance of responding to groundwater problems in locations where they are evident should be clear. While the 'global' warnings about groundwater overdraft may be welcome, they cannot in themselves point to any meaningful intervention or solution. These types of natural resource problems cannot be treated 'globally'. The issue is one of finding an appropriate physical and administrative level at which the 'common property' can be addressed successfully. The aggregate impact of such interventions will yield an improvement, but it will be incremental.

## Widening the Malthusian perspective

A Malthusian perspective on groundwater overabstraction leading towards a global food security crisis suggests courses of action designed to respond in an equally dramatic fashion to perceived threats to the resource base. In the groundwater case, this has led to a preoccupation with regulation, extraction control, water allocation and aquifer protection. While such activities will remain important, a lens that recognizes both variability and uncertainty requires a wider focus.

The global water management literature tends to emphasize the importance of complete institutional restructuring in the water sector and the development of 'comprehensive' and 'integrated' water management strategies. There are widespread calls for the development of integrated water rights systems, markets or other frameworks for allocating limited supplies and for the establishment of basin or aquifer management authorities with substantial regulatory powers. The focus is on management of the water resource base per se and not on the range of related services derived from groundwater use. This paper argues that a more 'clumsy' approach is essential to complement but not replace the standard management paradigm.

## The standard management paradigm

The standard 'integrated' approaches imply that systems are understood and can be 'managed' in a comprehensive manner. They presume an ability to identify and quantify the nature of interactions and to define the boundaries of systems clearly. They also presume that social institutions (rights systems, regulatory organizations, etc.) are present and contain sufficient capacity that management can be implemented in a planned and integrated manner. In the case of groundwater overdraft, assuming new supplies are not available, managing overdraft will require a reduction in use. In this situation, effective management requires:

- Sufficient technical capacity to quantify required use reductions, identify aquifer boundaries, and monitor the impact of management on aquifer conditions.
- Institutions for the equitable allocation of use reductions, which in turn generally depend on a rights or licensing system that is legally recognized, acceptable to users and enforceable.
- Effective mechanisms for coordinating the variety of agencies and users whose actions affect groundwater extraction and for planning the overall approach to management. In many situations, there are multiple departments and multiple users each with their own interests, perspectives and agendas.
- Effective mechanisms for educating users and for building social and political support for difficult management actions.

In many developed and developing countries, organizations and institutional frameworks capable of the above tasks are either weak or absent. Developing such frameworks can be a major task and must occur before any real management benefits will begin to flow. India has more than 20 million wells and a weak legislative framework for management. The task of well registration would require years of effort coupled with the introduction of a new legal framework. Therefore, effective implementation of the standard management paradigm is a long-term prospect at best.

## 'Clumsy' approaches

'Clumsy' approaches that respond to current constraints while not attempting to manage entire systems (and are therefore not fully 'integrated') are as important as more standard integrated management paradigms. Such adaptive approaches often work through lateral interventions (such as encouraging non-agricultural activities rather than attempting to control agricultural extraction) and can be designed to respond to variations in conditions and uncertainty. Because adaptive approaches do not require full understanding of resource dynamics and build off coping strategies that populations are already engaged in, such approaches may be able to produce 'results' more rapidly than can integrated management initiatives. In addition, they often do not require the introduction of new institutions (such as water rights) and may be able to minimize politically difficult decisions (such as extraction controls) in the short term. This is important because the institutional capacity and data essential for active integrated management of the resource base is absent in many countries and could take decades to develop. Adaptive strategies can provide a critical breathing space while the capacity for more direct integrated management is being built. Furthermore, adaptive strategies may prove more effective than integrated management in many cases because of their ability to respond rapidly to the process of social and environmental change that is ongoing in most parts of the world.

Clumsy approaches assume that tasks such as coordination are difficult and expensive. As a result, instead of attempting to develop a comprehensive integrated approach, the technique focuses on windows of opportunity that move groundwater conditions in the general direction required to address the problem. For example, when groundwater levels decline rapidly, some users may already be migrating in search of alternative sources of livelihood, others may be trying to focus on less water-intensive crops, and yet others may be trying to harvest more water. In this situation, interventions that help people to do what they are already trying to do on their own are likely to reduce pressure on the groundwater resource base. While better information and coordination could be helpful, the general direction in which change needs to go is clear and many practical things can be done that fit with that direction. Furthermore, many of the interventions do not require any coordination at high levels in order to be of some benefit at local levels (e.g. construction of local water harvesting structures or encouragement of non-agricultural activities). However, this may be the best case. Farmers pumping groundwater may not be water sensitive if well adjusted to a crop with high water consumption. The example of sugar cane is instructive as it is one of the few crops for which companies will offer credit to smallholders from the day of planting. These types of inelastic behaviour inhabit the mosaic of water use in rural settings and need to be taken along with the more positive opportunities for enhancing adaptive behaviour with regard to groundwater.

### The significance of adaptive approaches in groundwater management

As Moench (1999) argues, four elements appear central to the development of adaptive responses to groundwater problems under conditions of variability and uncertainty:

- systemic perspectives;
- constraint analysis;
- identification of an appropriate physical and administrative level at which the 'common property' can be addressed successfully (i.e. the response needs to be context reflective);
- social auditors (the engine of change).

Each of the four elements functions more effectively in a situation where whatever data exist are accessible and can be used by different groups within civil society. Thus, although this does not remove uncertainty, data access is also an important condition.

There is a core distinction between using a systemic perspective in developing water management responses and attempting to develop comprehensive integrated responses. A systemic perspective recognizes both the importance of interactions between systems and the limitations of knowledge regarding those interactions. It also emphasizes scale issues. Aquifers or watersheds are not discrete units but operate rather as systems within systems. The scale at which management needs to occur and the type of management both depend on the scale of system processes and the nature of system interactions. Because these interactions are complex, a systemic perspective leads to a focus on those factors that appear at any given time to be the 'drivers' of underlying problems. Unlike the 'comprehensive integrated' terminology, a systemic perspective does not convey the impression that all driving factors are known and incorporated in the approach proposed. The distinction is subtle but important. Describing an approach as being comprehensive and integrated gives readers (including those involved in critical policy and decision-making roles) the impression that all relevant factors have been reviewed 'comprehensively' and 'integrated' effectively into the plan of action. In most situations, this is not possible. Gaps in data, basic scientific knowledge, social dynamics and many other factors make comprehensive integration impossible. In contrast, a systemic perspective draws attention to the variety of factors influencing water management needs and options in a given situation but without claiming an unrealistic ability to be either integrated or comprehensive. As a result, the perspective should focus attention on the factors that analysts view as most important while leaving the door open to changing management approaches as knowledge improves and conditions evolve.

As social dynamics change, so do the factors driving resource conditions. A systemic perspective is intended to encourage recognition of dynamic change processes and to encourage adaptation to them. Because a systemic perspective does not generate the impression that all factors are being addressed, it encourages rapid responses to changes and new information. A systemic perspective is also intended to situate responses within the wider context while de-emphasizing the importance of comprehensive integrated management planning. In contrast, integrated approaches are based on the assumption that all the important interactions can be identified and incorporated into the management approach. Because it is generally impossible to achieve this in practice, approaches based on comprehensive integrated planning generate a false sense of accuracy and security regarding the directions management should take and the results it should have. In addition, such management planning often absorbs considerable resources and produces products that are obsolete before they are implemented.

Constraint analysis is the servant of a systemic perspective. Instead of attempting to manage a system as a whole, the approach is to focus on the constraints that affect key values. In the groundwater case, falling or rising water levels in regional aquifers are often the primary constraint affecting agriculture, the environment, regional economies and other values. Constraints also take the form of surprises (low-probability catastrophic events) such as the occurrence of regional droughts in areas affected by groundwater overabstraction. At a local level, these surprises may have more important implications for food security than the more gradual, long-term process of groundwater overabstraction or water-level decline. Compared with 'integrated' analysis, focusing on constraints and surprises may provide more insights into the problems people face and potential courses of action to address them.

Adaptive or context-reflective responses follow from the combination of a systemic perspective and constraint analysis. Because the lens through which groundwater problems are viewed emphasizes variability, uncertainty and the lack of key systems information necessary for comprehensive management, adaptation needs to be a core component of any strategy. The argument behind this is not simply that the data needed to understand hydrological systems are lacking. It also reflects the role data play in developing social consensus around management strategies. Where overabstraction is present, management of the resource base would generally require large reductions in use and the allocation of available supplies through a rights system or other similar mechanism. For example, in the case of aquifers in northern Gujarat, India, 1976 estimates suggested that extraction would need to be reduced by 25 percent to reach sustainable levels (United Nations Development Programme, 1976). Percentage reductions would now need to be considerably higher. In most cases, building sufficient social consensus for this type of reduction to work would depend on a long-term process through which understanding of aquifer dynamics is generated and disseminated to the key actors, i.e. the well owners whose behaviour determines overabstraction conditions. In locations where this has occurred (e.g. the western United States of America), it has generally involved many decades of data collection and social debate among a limited set of highly educated players before action could be taken. Even so, the results in such cases have been mixed. Groundwater user associations in Mexico have reportedly agreed to substantial reductions in extraction much more rapidly, but the results are yet to appear. Expecting this approach to be effective on a short time scale under conditions in many developing countries appears unrealistic. While not suggesting that efforts to develop management capacity should be reduced, it is equally important to respond and adapt to constraints. In other words, building the capacity of society to adapt to water constraints is as important as the capacity to modify or manage water systems themselves.

The philosophy underlying an adaptive or context-reflective approach to groundwater problems emphasizes the development of responses based on the specific constraints emerging in a local situation. Rather than a uniform set of 'best management practices', it recognizes that management needs and capabilities vary considerably. It emphasizes adaptation in two ways: (i) adaptation of water management to the needs and opportunities present in local contexts; and (ii) adaptation of society to the constraints emerging from water-related limitations. The broad focus of the lens is evident in the latter. In most cases, governments attempt to respond to water problems through water-related interventions. Social dynamics (migration, changing regional economic systems, etc.) are not generally part of the formal management package. However, they are often the main response individuals are making as they face water-related constraints. In addition, while they affect water use patterns and management options, they are often driven by factors other than water. Adaptive approaches would attempt to identify these types of dynamics and build the types of governmental and other interventions that would support adaptation rather than controlling groundwater use. In this sense, adaptive responses constitute a form of strategic planning that is by nature dynamic, iterative and responsive to uncertainty. It suggests focusing management in the short term on priority aquifers where capacity and incentives are present at the grassroots level, possibly through networks of cooperative institutions, rather than relying on national institutions which often require long-term effort to develop. This does not diminish the importance of longer-term aquifer research and planning and the development of national institutions to undertake these tasks. These are still essential to framing management responses (Burke, 1996) and avoiding anarchic and potentially destructive development over the longer-term.

On one level, the proposed approach reflects a pragmatic response to power relations and dynamics within society as it is often not politically or socially viable to control extraction and, as a result, adaptation is essential. However, on another level, the approach recognizes that users are often 'adapting' for reasons that reflect a larger set of aspirations and goals in addition to water constraints. The small farmers who decide to pump as much as possible in order to educate their children are probably behaving rationally. Forcing them to reduce extraction could lock them in a vicious cycle of marginal agriculture and poverty. In contrast, helping them to educate their children and move out of agriculture could create a more sustainable livelihood for their families and ultimately lead to reductions in groundwater extraction. Thus, adaptation may open new perspectives that support sustainable livelihoods and poverty alleviation rather than highlighting perceived conflicts between resource sustainability and current use patterns.

It is important to recognize that adaptive responses do not eliminate the importance of attempts to develop groundwater management systems. However, they do recognize that the development of management systems is generally a long-term process and often will not provide an effective avenue for addressing the impacts of groundwater overabstraction within the near future. Furthermore, adaptive strategies that move communities away from groundwater dependence could provide the breathing space and social conditions necessary for institutions capable of managing groundwater to evolve. For example, groundwater management in the United States of America is often effectively in the hands of a few tens to hundreds of actors (Blomquist, 1992). This enables negotiations to occur between small groups of stakeholders representing interest groups rather than the thousands of individual well owners often found in locations such as India. Adaptive strategies that move many individuals away from groundwater dependency could help to create the conditions under which direct use of groundwater becomes consolidated in a way that encourages long-term planning and management.

The term 'social auditors' represents a final component that is essential for the evolution of social capacity to respond to water constraints. Social auditors are: "*the 'watch dog' social activists as well as various organs of the state that are responsible for assuring appropriate justice. They are not users or managers, and their concerns often stem from different callings - those of equity, sustainability and fair play. Linear policy models that account for the users at the bottom and the managers at the top are often at a loss when these actors enter the fray - often in the event that contradictions emerge between the avowed objectives of management and its practice. Except for extreme cases of bureaucratic rigidity, social auditors from the activist mould and from within the government do often work together to assure proper functioning by the concerned water bureaucracy. Even as the managers of a department may advocate hierarchical administrative approaches to water management, significant sections of the state machinery, including the judiciary and units of local governance, often assert themselves in upholding points of equity, democratic process and social justice.*" (Moench *et al.*, 1999).

Social auditors serve as catalysts that highlight emerging water problems and their social impact. They also focus social attention on potential solutions. In many cases, they represent a critical social element balancing the power of government bureaucracies or giving voice to socially marginal sections of society. Their ability to play this role is enhanced by access to information and technical-analytical capabilities.

## Implications for groundwater management

The results of this research point to the need to rethink groundwater management for food security. The paper contains numerous contradictions. While acknowledging the fundamental importance of emerging groundwater problems, the research led in an unanticipated direction. Instead of facts and documentation accumulating towards an increasingly clear and well-founded perspective regarding the impact of groundwater overdraft on food security, variability and scientific unknowns have tended to accumulate. However, groundwater overdraft may be justified in economic terms where the immediate drawdown externalities are negligible and the long-term benefits derived from its use allow sustainable substitutes to groundwater as a factor of production to be developed (Schiffler, 1998). It is not always possible to state categorically that groundwater overdraft is inherently 'bad'.

The demand for groundwater irrigation is diverse. It ranges from smallholder irrigation in remote locations to meet local needs (typical of Sahelian irrigation from basement complex aquifers) to large-scale commercial irrigation to feed national markets (wheat irrigation based on highly productive dolomite aquifers in Zambia). Each type of demand and style of irrigation has its own, usually highly individual, management strategies that are conditioned by specific groundwater dynamics. Some of these patterns of demand will change as populations migrate, subsidy structures change and alternative opportunities arise. These changes will need to be sensed before implementing new policy recommendations or advocacy.

Overabstraction, quality and other groundwater problems are severe in many parts of the world. However, they need to be set against the significance of reliable water services in maintaining global food security. At present, irrigated agriculture in developing countries accounts for only 20 percent of the arable lands but contributes nearly 60 percent of cereal production and 40 percent of total crop production (FAO, 2002a). However, groundwater generally produces substantially higher yields than areas irrigated from surface sources and, in locations such as Spain, can generate economic returns up to five times those found in surface irrigated areas (Hernandez-Mora *et al.*, 1999). Because groundwater plays a major role in achieving high yields, particularly where demand for high-value crops is effective, emerging threats to the resource base do have important food security implications at local and possibly global levels.

However, the accessible groundwater data are insufficient to quantify the areas where food security implications may be severe. Furthermore, emerging groundwater problems are important regardless of their implications for global food security. Effective approaches to addressing these problems are essential. However, the same challenges that frustrate attempts to estimate the impact of groundwater overabstraction on food security also complicate the applicability of standard management models. The contradictions that are evident in this paper are also inherent in the groundwater story itself:

- Problems are emerging but there is uncertainty regarding their nature, extent and implications.
- Management needs are evident but the viability of management is unclear in many situations.
- Groundwater problems create significant inequities in access to water for poor and marginal users but most standard management solutions would in practice exacerbate such inequities.
- Solutions that ensure access to groundwater for all exist but they are likely to lead to the depletion of the resource base and ultimately increase inequity.
- Solutions to overabstraction that build off inequity by helping those displaced by overabstraction to develop alternative livelihoods may ultimately be the most equitable if they ease pressure on the resource base and enable people to move out of poverty.

On closer examination, the above contradictions may prove more apparent than real. However, they point to a major need to rethink groundwater use for irrigated agriculture. Better understanding is needed regarding the implications of emerging problems and the best points of leverage for addressing them. Better understanding is also needed with respect to the coping strategies that people follow and how attempts to address groundwater problems can build off existing trends in society rather than attempt to work against them.

# Chapter 5
# Recommendations

The results of the analysis in this paper suggest two broad avenues for future work. The first involves the development of national research programmes to gather representative groundwater data directly in countries and regions where dependence upon groundwater is high. This type of overview is essential in order to develop more informed assessments of the implications of groundwater conditions for food security and to develop scientifically founded courses of action for managing the resource base. The second avenue for work focuses on the development of adaptive responses to water problems and policy approaches that reflect and respond to uncertainty, change and the absence of real understanding of systems and their interactions. Inherent limitations in the nature of hydrogeological information in conjunction with the social and institutional changes occurring in many parts of the world make this second avenue of work at least as important as the first. A key part of this should include the development of early-warning indicators that help identify where groundwater problems in conjunction with other local factors indicate the emergence of local or regional food security problems.

In addition to these two broad areas, the analysis suggests a variety of key points of leverage for technical assistance organizations to assist in developing effective responses to emerging groundwater problems. These points of leverage are listed below in a particular sequence for specific reasons. Effective responses to emerging groundwater problems are essential. However, this paper argues that the viability of traditional integrated management approaches is limited by a wide array of data, technical, social and political factors. As a result, society needs to proceed on two equally important courses. One course focuses on adaptation to problems and on building the basic understanding from which both adaptive and management responses must flow. The second course, of equal but not more importance than the first, involves a continuation of current efforts towards the integrated management of groundwater. However, instead of forming the dominant strategy, this course focuses on strategic locations where the technical, political and social viability of management appears strong. The points of leverage involve a combination of: (i) efforts to rethink and implement new approaches to groundwater with particular emphasis on adaptive strategies; (ii) basic research on the nature of groundwater problems and the larger social context in which they are situated; (iii) continued attention to the basic science, data and information generation activities that provide a foundation for all understanding of groundwater issues; (iv) continued efforts to build integrated groundwater management capacity in strategic locations where management appears viable in the short to intermediate term; (v) initiatives to build the foundations for management in more difficult locations where results can only be expected in the long term; and (vi) efforts to harvest and disseminate lessons from the growing global experience with groundwater and its management. The potential activities and needs in each of these areas are discussed below, but in approaching these points of leverage, constant attention will need to be given to the level at which information is being compiled and management applied. These levels have to be consistent with the scale of the aquifer systems and the 'administration' of the common property.

## Rethinking the approach to groundwater management

A rethink of approaches to groundwater management with a focus on adaptive approaches is a principal recommendation of this report. Standard management approaches depend heavily on the presence of basic data and on institutional capacities for regulation, scientific research, etc. at levels that may be incompatible with the problems at hand. In addition, these capacities are absent or weak in many countries. Because such capacities and data often require decades to develop, alternative approaches are essential in order to address the types of problems that are now emerging in many regions that are dependent on groundwater. Furthermore, the research suggests that strategies that build off existing trends within society or help populations to adapt may be as effective as strategies that attempt to manage the groundwater resource base directly. Research to clarify existing coping mechanisms and to identify and test the viability of adaptive strategies could represent a major starting point for an initiative to rethink groundwater. The development of criteria suggesting where traditional forms of groundwater management may or may not be possible is also a key area for work. This could be of critical importance to governments and other actors seeking to identify locations where different approaches are likely to prove viable. Finally, the development of indices to highlight where local or regional threats to food security exist and to provide early warning of emerging food security problems is important. Such indices should include extensive information on groundwater. In particular, areas of groundwater depletion should be highlighted as locations where food security risks may be high.

## Basic research

Basic research on groundwater is of fundamental importance to any attempt to manage groundwater or respond to the types of problems emerging in many areas. As noted in the case of India, the implications for water access of seasonal water-level fluctuations (whether natural or related to extraction) may be at least as important as the extent of overabstraction. Large-magnitude fluctuations may be particularly important in hard-rock regions where storage is low but human dependence is high

In addition to specific environments, further research to identify techniques for the rapid and accurate evaluation of water-balance components under developing-country conditions is important. Order-of-magnitude estimates for extraction, recharge, evapotranspiration, etc. are unavailable in many regions. Research that would help to 'tighten' and improve water-balance models with the types of data typically available in developing countries is important.

Beyond groundwater per se, it appears important to focus further research on the changing social context in which overabstraction problems are emerging. Understanding the implications of groundwater overabstraction for food security and livelihood sustainability requires detailed understanding of how rural agricultural societies are evolving and of the coping strategies they have developed to deal with water scarcity. Whether or not people are actually able to move their livelihoods away from agriculture into other productive strategies is fundamental to understanding the impact that overabstraction may have on them. This research is essential in order to determine whether adaptive strategies can improve livelihoods while increasing the sustainability of basic groundwater resources. It is also essential in order to identify points of leverage where governments or other organizations could assist rural populations in adapting to emerging water problems.

### Groundwater monitoring and data collection, dissemination and access

Many governments and other actors attach a low priority to collecting basic data on resource conditions. However, such long-term data covering all key elements of the hydrological cycle including groundwater fluctuations and water-level trends are essential as a basis for management and for evaluating the implications of changes in use. The ability to locate representative monitoring wells and boreholes is essential. In addition, models based on such data can play a central role as 'negotiating texts' where conflicts over resources or their management emerge within society. In the absence of such data, the parties have little basis for reaching agreement on the actual nature of groundwater systems. As a result, debates over management have little hope of reaching closure. Because data provide the foundation for social agreements regarding how aquifer systems work or the actual amount of water available, they can serve as a key tool of conflict resolution. Therefore, continued support for basic data collection and groundwater evaluation is justified on both scientific and social process grounds.

Data access is probably the most important factor determining the ability of social auditors (e.g. NGOs and other civil society actors) to press governments and society as a whole to address emerging problems and their social or environmental impacts. Therefore, activities that support data dissemination remain a key point for action. The FAO Aquastat database provides a national breakdown of groundwater-dependent irrigation (accessible at: http://www.fao.org/ag/AGL/aglw/Aquastat/aquastat.htm). However, attempts to refine this breakdown will encounter the data problems demonstrated above in the case of India.

### Integrated management in strategic locations

Together with the need to rethink groundwater and identify new strategies to address emerging problems, continued efforts to implement integrated groundwater management using more standard regulatory and economic approaches are equally important in locations where such approaches appear viable. Because standard management approaches tend to require substantial technical support and often involve politically or economically difficult decisions, success may depend on focusing management initiatives in areas of particular strategic importance. For example, aquifers that serve as the primary source of freshwater supply for urban areas or support critical environmental values may represent strategic locations on which to focus management efforts. In most countries, such aquifers represent a small fraction of total groundwater use. They are also likely to involve uses where it is relatively easy to generate broad consensus within society regarding the importance of management and aquifer protection. As a result, approaches that focus management on such strategic locations are more likely to be successful than efforts to manage groundwater throughout broad regions.

FAO could contribute significantly to the development of management capacity by developing criteria that would assist in prioritizing management areas and identifying locations where protection would have particular strategic importance.

### Laying the foundations for management in complex locations

Adaptive strategies are of equal importance and complementary to more standard groundwater management approaches. In many cases, they can provide the breathing space necessary to develop the institutions and information essential for more focused management. Thus, it is important to continue to lay the foundations for direct groundwater management even where it

**Plate 8**
**Developing groundwater to irrigate horticultural crops with drip. Batinah plain, Oman**
*[J.J. Burke]*

may not produce results in the short to intermediate term. Therefore, continued FAO support for basic groundwater data collection, the development of legal frameworks to enable management and the development of supporting organizations is important.

## Disseminating global lessons

A final key point of leverage for UN-system agencies lies in the global perspective they can bring to groundwater based on actual national data sets. Governments and communities in many parts of the world are trying different approaches to groundwater monitoring, analysis and management. Harvesting and disseminating the lessons from these initiatives could serve as a catalyst for the development of approaches that are effective even in the most difficult locations. As a result, activities that support the harvesting and dissemination of instances of adaptive groundwater management (simply to show what happens) will continue to be an important activity for UN agencies involved in groundwater management. The actual experience of groundwater management, or the lack of it, needs to be charted if real responses are to be effective (Plate 8).

# Bibliography


**Amarasinghe, U.A., Mutuwatta, L. & Sakthivadivel, R.** 1999. *Water scarcity variations within a country: a case study of Sri Lanka*. Colombo, International Water Management Institute. 40 pp.

**Barraque, B.** 1998. Groundwater management in Europe: regulatory, organisational and institutional change. *In* T. Hilding-Rydevik & I. Johansson, eds. *How to cope with degrading groundwater quality in Europe*, pp. 74-101. Stockholm.

**Basak, P., Lekha, K.R. & Prasad, P.M.R.** 1993. *Flow through fractured rocks. A field correlation between fracture parameters and yield.* Paper presented at the Workshop on Water Management: India's Groundwater Challenge. Ahmedabad, India.

**Blomquist, W.** 1992. *Dividing the waters: governing groundwater in southern California*. San Francisco, USA, Institute for Contemporary Studies.

**Brown, L. & Halweil, B.** 1998. China's water shortage could shake world food security. *Wld. Watch Mag.*

**Brown, L.** 1999. Feeding nine billion. *In* L. Starke, ed. *State of the world 1999*, pp. 116-132. New York, USA, W.W. Norton & Company.

**Burke, J.J.** 1996. Hydrogeological provinces in central Sudan: morphostructural and hydrogeomorphological controls. *In* T. Brown, ed. *Groundwater and geomorphology*. Chichester, UK, Wiley.

**Burke, J.J.** 2000. Land and water systems: managing the hydrologic risk. *Nat. Res. For.,* 24: 123-136.

**Burke, J.J.** 2002. Groundwater for irrigation: productivity gains and the need to manage hydro-environmental risk. *In* R. Llamas & E. Custodio, eds. *Intensive use of groundwater challenges and opportunities,* 478 pp. Abingdon, UK, Balkema.

**Burke, J.J. and M. Moench.** 2000. *Groundwater and society: resources, tensions, opportunities*. New York, United Nations. 170 pp.

**Central Ground Water Board (CGWB).** 1995. *Groundwater resources of India*. Faridabad, India.

**Chambers, R., Saxena, N.C. & Shah, T.,** eds. 1989. *To the hands of the poor, water and trees*. New Delhi, Oxford and IBH Publishing Co.

**Dains, S.R. & Pawar, J.R.** 1987. *Economic returns to irrigation in India*. New Delhi, Report prepared by SDR Research Groups Inc. for the US Agency for International Development.

**Dreze J., Sen, A. & Hussain, A.,** eds. 1995. *The political economy of hunger*. New Delhi, Oxford University Press.

**Eckholm, E.P.** 1976. *Losing ground: environmental stress and world food prospects*. New York, USA, W.W. Norton & Company, Inc.

**FAO.** 1994. *Management of groundwater resources in China*, by S. Lunzhang. Rome.

**FAO.** 1996. *Food, agriculture and food security: developments since the World Food Conference and prospects*. Rome.

**FAO.** 2002a. *World agriculture: towards 2015/30*. Summary Report. Rome.

**FAO.** 2002b. *Water and fertilizers use in selected countries*. Discussion Paper. Land and Water Development Division, FAO. Rome.

**Gleick, P. & Nash, L.** 1991. *The societal and environmental costs of the continuing California drought.* Oakland, USA, Pacific Institute for Studies in Development, Environment and Security.

**Goldman, M.R.** 1988. *Cultivating hot peppers and water crisis in India's desert.* Santa·Cruz, USA, University of California, Sociology Board, Merril College.

**Ground Water Resource Estimation Committee (GWREC).** 1997. *Ground water resource estimation methodology - 1997.* New Delhi, Ministry of Water Resources.

**Hernandez-Mora, N., Llamas, R. & Martinez-Cortina, L.** 1999. *Misconceptions in aquifer overexploitation: implications for water policy in southern Europe.* Workshop SAGA (FAIR-CT97-3673), Milan, Italy, Fondazione Enni Enrico Mattei.

**Ives, J.D. & Messerli, B.** 1989. *The Himalayan dilemma: reconciling development and conservation.* New York, USA, Routledge.

**IWHR.** 1998. *Water availability and projection on water demand in the future.* Beijing, Institute of Water Resources and Hydropower Research.

**Kahnert, F. & Levine, G.** 1989. *Key findings, recommendations, and summary. Groundwater irrigation and the rural poor: options for development in the Gangetic basin.* Washington, DC, The World Bank.

**Ladha, J.K., Fischer, K.S., Hossain, M., Hobbs, P.R. & Hardy, B.** 2000. *Improving the productivity and sustainability of rice-wheat systems of the Indo-Gangetic plains: a synthesis of NARS-IRRI partnership research.* IRRI Discussion Paper No. 40. Los Baños, Philippines, International Rice Research Institute.

**Meinzen-Dick, R.** 1996. *Groundwater markets in Pakistan: participation and productivity.* Washington, DC, International Food Policy Research Institute.

**Ministry of Water Resources.** 2001. *China agenda for water sector strategy for north China.* Volume 2: Main Report April 2, 2001 Report No. 22040-CHA.

**Moench, M.** 1991. *Sustainability, efficiency, and equity in ground water development: issues in the western U.S. and India.* Oakland, USA, Pacific Institute for Studies in Environment, Development and Security.

**Moench, M.** 1992. Drawing down the buffer. *Econ. Pol. Week.,* XXVII(13): A-7-A-14.

**Moench, M.** 1994. Hydrology under central planning: groundwater in India. *Wat. Nep.,* 4(1): 98-112.

**Moench, M.** 1996. *Groundwater policy: issues and alternatives in India.* Colombo, International Irrigation Management Institute.

**Moench, M.** 1998. *Groundwater: the underlying resource, themes in groundwater management for the 21st century.* New York, United Nations, Department for Social and Economic Affairs and Institute for Social and Environmental Transition.

**Moench, M.** 1999. Addressing constraints in complex systems. *In* M. Moench, E. Caspari & A. Dixit, eds. *Rethinking the mosaic: investigations into local water management.* Kathmandu, The Collaborative Project on Local Water Management.

**Moench, M., Caspari, E. & Dixit, A.,** eds. 1999. *Rethinking the mosaic: investigations into local water management.* Kathmandu, Nepal Water Conservation Foundation and the Institute for Social and Environmental Transition.

**Narasimhan, T.N.** 1990. Groundwater in the peninsular Indian shield: a framework for rational assessment. *J. Geo. Soc. India,* 36: 353-363.

**Ostrom, E.** 1993. Design principles in long-enduring irrigation institutions. *Wat. Res. Res.,* 29(7): 1907-1912.

**Perry, C.J. & Narayanamurthy, S.G.** 1998. *Farmer response to rationed and uncertain irrigation supplies.* Colombo, International Water Management Institute.

**Pinstrup-Anderson, P., Pandya-Lorch, R. & Rosegrant, M.W.** 1999. *World food prospects: critical issues for the early twenty-first century*. Washington, DC, International Food Policy Research Institute. 32 pp.

**Pisharote, P.R.** 1992. *Rainfall regime of Kutch District, Gujarat State, India*. Workshop on Drought Management, Mandvi, Kutch, India.

**Postel, S.** 1999. *Pillar of sand: can the irrigation miracle last?* New York, USA, Norton.

**Rathore, M.S. & Mathur, R.M.** 1999. Local strategies for water management and conservation: a study of Shekhawati basin, Rajasthan. In M. Moench, E. Caspari & A. Dixit, eds. *Rethinking the mosaic: investigations into local water management*. Kathmandu, Nepal Water Conservation Foundation and Institute for Social and Environmental Transition.

**Reisner, M.** 1986. *Cadillac desert*. New York, Penguin.

**Repetto, R.** 1994. *The "second India" revisited: population, poverty, and environmental stress over two decades*. Washington, DC, World Resources Institute.

**Rosegrant, M.W. & Ringler, C.** 1999. *Impact of food security and rural development of reallocating water from agriculture*. Washington, DC, International Food Policy Research Institute.

**Rosenzweig, C. & Hillel, D.** 1995. Potential impacts of climate change on agriculture and food supply. *Consequences,* 1(2): 12.

**SASS/OSS.** 2001. *Acquisition, organisation and analysis of the SASS hydrological data*. Tunis, OSS.

**Schiffler, M.** 1998. *The economics of groundwater management in arid countries*. GDI Book Series II. London, Frank Cass. 394 pp.

**Seckler, D. & Amarasinghe, U.A.** 1999. *Water supply and demand, 1995 to 2025: water scarcity and major issues,* Chapter 3. From IWMI Web Site. Colombo.

**Seckler, D., Amarasinghe, U.A., Molden, D., de Silva, R & Barker, R.** 1998. *World water demand and supply, 1990 to 2025: scenarios and issues*. Colombo, International Water Management Institute. 52 pp.

**Seckler, D., Barker, R. & Amarasinghe, U.A.** 1999. Water scarcity in the twenty-first century. *Int. J. Wat. Res. Dev.,* 15(1/2): 29-42.

**Sen, A.** 1999. *Poverty and famines*. Delhi, India, Oxford University Press.

**Shah, T.** 1993. *Groundwater markets and irrigation development: political economy and practical policy*. Mumbai, India, Oxford University Press.

**Shah, T., Molden, D., Sakthivadivel, R. & Seckler, D.** 2000. *The global groundwater situation: overview of opportunities and challenges*. Colombo, International Water Management Institute. 30 pp.

**Thompson, M., Warburton, M. & Hatley, T.** 1986. *Uncertainty on a Himalayan scale*. London, Milton Ash Editions.

**Tsur, Y.** 1990. The stabilization role of groundwater when surface water supplies are uncertain: the implications for groundwater development. *Wat. Res. Res.,* 26(5): 811-818.

**Tsur, Y.** 1993. *The economics of conjunctive ground and surface water irrigation systems: basic principles and empirical evidence from southern California 22*. USA, Department of Agricultural and Applied Economics, University of Minnesota.

**United Nations Development Programme**. 1976. *Ground-water surveys in Rajasthan and Gujarat*.

**United Nations.** 1997. *Comprehensive assessment of the freshwater resources of the world*. Commission on Sustainable Development.

**United Nations.** 1999. *Perspectives on integrated basin management for a policy study on the Huaihe river basin, China*. Report of the Department of Economic and Social Affairs.

**FAO TECHNICAL PAPERS**

WATER REPORTS



Availability: January 2003

Ar – Arabic
C – Chinese
E – English
F – French
P – Portuguese
S – Spanish

Multil – Multilingual
* Out of print
** In preparation